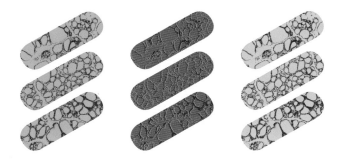 OXFORD BIOLOGY PRIMERS

Discover more in the series at
www.oxfordtextbooks.co.uk/obp

Published in partnership with the Royal Society of Biology

MAMMALIAN SYNTHETIC BIOLOGY

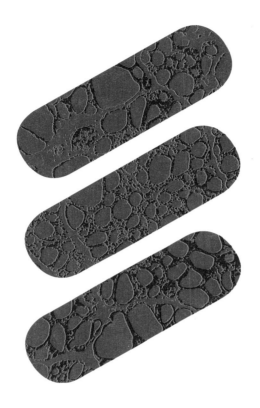

MAMMALIAN
SYNTHETIC BIOLOGY

Edited by Jamie A. Davies
Subject Editor: Paul S. Freemont

OXFORD
UNIVERSITY PRESS

Royal Society of
Biology

OXFORD
UNIVERSITY PRESS

Great Clarendon Street, Oxford, OX2 6DP,
United Kingdom

Oxford University Press is a department of the University of Oxford.
It furthers the University's objective of excellence in research, scholarship,
and education by publishing worldwide. Oxford is a registered trade mark of
Oxford University Press in the UK and in certain other countries

Published in the United States of America by Oxford University Press
198 Madison Avenue, New York, NY 10016, United States of America

British Library Cataloguing in Publication Data
Data available

Library of Congress Control Number: 2019947539

ISBN 978-0-19-884154-8

Printed in Great Britain by
Bell & Bain Ltd., Glasgow

To our colleagues and students in Edinburgh's *SynthSys* community, who make doing and teaching synthetic biology such an interesting experience for all of us.

PREFACE

Welcome to the Oxford Biology Primers

There has never been a more exciting time to be a biologist. Not only do we understand more about the biological world than ever before, but we're using that understanding in ever more creative and valuable ways.

Our understanding of the way our genes work is being used to explore new ways to treat disease; our understanding of ecosystems is being used to explore more effective ways to protect the diversity of life on Earth; and our understanding of plant science is being used to explore more sustainable ways to feed a growing human population.

The repeated use of the word 'explore' here is no accident. The study of biology is, at heart, an exploration. We have written the Oxford Biology Primers to encourage you to explore biology for yourself—to find out more about what scientists at the cutting edge of the subject are researching, and the biological problems they're trying to solve.

Throughout the series, we use a range of features to help you see topics from different perspectives.

Scientific approach panels help you understand a little more about 'how we know what we know'—that is, the research that has been carried out to reveal our current understanding of the science described in the text, and the methods and approaches scientists have used when carrying out that research.

Case studies explore how a particular concept is relevant to our everyday life, or provide an intimate picture of one aspect of the science described.

The bigger picture panels help you think about some of the issues and challenges associated with the topic under discussion—for example, ethical considerations, or wider impacts on society.

More than anything, however, we hope this series will reveal to you, its readers, that biology is awe-inspiring, both in its variety and its intricacy, and will drive you forward to explore the subject further for yourself.

ABOUT THE AUTHORS

Jamie Billington MSc

Jamie completed a BA in Natural Sciences and then an MSc in Systems Biology at the University of Cambridge, graduating in 2016. During his masters, he worked on a project to develop new genetic toggle switches in *Escherichia coli*. Since then, he has been pursuing a PhD at the University of Edinburgh in Professor Susan Rosser's lab, where he uses synthetic biology tools to engineer mammalian cells for bioproduction. His research focuses include controlling chromatin modifications on synthetic genes and identifying insulator sequences.

Professor Jamie A. Davies PhD, FRSM, FRSB, FRSA, PFHEA, FAS, MIEEE

Jamie read Natural Sciences at Cambridge University, where he went on to gain a PhD in Developmental Neurobiology. He worked as a post-doctoral fellow in Cancer Research Campaign laboratories at the University of Southampton and the University of Manchester, then set up his own laboratory at the University of Edinburgh, where he is now Professor of Experimental Anatomy. His laboratory explores basic mechanisms of embryonic development and applies them to the problem of making new tissues and organs from stem cells, and tests hypotheses about developmental mechanisms by recreating them in naive cells using synthetic biology. He is the author of around 200 research papers and author/editor of eleven books, has given public presentations live and on radio programmes, and is a teacher of these topics at undergraduate and postgraduate levels.

Professor Alistair Elfick PhD, FRSB

Alistair is Head of the Institute for Bioengineering and Deputy Director of the UK Centre for Mammalian Synthetic Biology at the University of Edinburgh. Researching at the boundary of life and the inanimate, Alistair seeks to advance humankind's capacity to maintain health by sympathetically enhancing our ability to recognize disease and heal. Alistair gained degrees in Mechanical Engineering and Biomedical Engineering at the University of Durham, UK. He won both a Fulbright Commission, Distinguished Scholar's Award, and a Royal Academy of Engineering, Global Research Award to experience biomedical engineering research at the University of California, Berkeley. Having enjoyed too much Californian sunshine, Alistair returned to his hometown of Edinburgh in 2004 as recipient of an EPSRC Advanced Research Fellowship, taking his Personal Chair in Synthetic Biological Engineering in 2012.

Anna Mastela MSc

Anna completed an MSc in Biotechnology at the University of Wroclaw, Poland. She is currently undertaking a PhD at the University of Edinburgh in collaboration with UCB Celltech, with her area of research focusing on the application of combined omics and synthetic biology approaches to enhance biologics production in CHO cells. She is a member of the Rosser Laboratory based in the UK Centre for Mammalian Synthetic Biology Research. Prior to starting her PhD, Anna worked as a Development Scientist at the Centre for Process Innovation Ltd. Her role was centred on improving upstream production of biologics.

Dr Leonard J. Nelson PhD

Lenny is a senior researcher in biological engineering at the Institute for Bioengineering at the University of Edinburgh. He was born in Edinburgh, completing a BSc (Honours) in Physiology at the University of Edinburgh in 1990. After obtaining a Diploma in Education and teaching chemistry and biology, he pursued his research career by completing a PhD in 2002 at the University of Edinburgh Faculty of Medicine (Hepatology) in the Liver Cell Biology Laboratory, working on liver tissue engineering approaches towards the development of bioartificial liver systems. Currently, his research is focused on developing human liver tissue (two/three-dimensional) organotypic models, mechanisms underlying liver disease utilizing *in vitro* human disease models (drug-induced liver injury; fatty liver disease; cholangiocarcinoma), study of the liver–brain axis in Alzheimer's disease; and synthetic biological engineering approaches for pharmaceutical applications.

David Obree BA (Hons), MSc, BDS, MFGDP RCS (Eng)

David is a medical ethics teacher at the University of Edinburgh Medical School and is currently the Archie Duncan Fellow in Medical Ethics and Medical Education at the Usher Institute. With degrees in dentistry, drama, and medical ethics, David has developed a special interest in how the humanities can help us make sense of complex moral dilemmas, such as those posed by the development of synthetic biology. His favourite philosophers are Bertrand Russell and Monty Python.

Professor Steven M. Pollard PhD

Steve carried out his PhD at the MRC National Institute for Medical Research (NIMR) in the Division of Developmental Biology. He trained as a postdoctoral scientist at the University of Edinburgh and University of Cambridge before establishing his own independent research team at University College London in 2009. He subsequently moved to the MRC Centre for Regenerative Medicine and Edinburgh Cancer Research Centre at the University of Edinburgh in 2013. He was appointed to Chair of Stem Cell and Cancer Biology in 2017 and plays an active role in the UK Centre for Mammalian Synthetic Biology. He has held a JB and Millicent Kaye Fellowship (Christ's College, Cambridge), an Alex Bolt Group Leader Fellowship (University College London), and is currently supported by the prestigious Cancer Research UK Senior Research Fellowship. His laboratory explores the molecular and cellular mechanisms that regulate neural stem self-renewal and differentiation, and how these operate in the context of human brain tumours. He is exploiting synthetic transcription factors and chromatin editors for programming and reprogramming mammalian cell fate.

Professor Susan J. Rosser PhD

Susan is currently Professor of Synthetic Biology at the University of Edinburgh, Director of the Edinburgh Mammalian Synthetic Biology Research Centre, and Co-director of the Edinburgh Genome Foundry. She has worked in the Institute of Biotechnology at the University of Cambridge and at the Institute of Molecular, Cell and Systems Biology at the University of Glasgow. Susan is a member of the Scottish Industrial Biotechnology Development Group and the Scottish Science Advisory Council, Scotland's highest-level science advisory body, providing independent advice and recommendations on science strategy, policy, and priorities to the Scottish Government.

CONTENTS

Abbreviations   xv

1   An introduction to mammalian synthetic biology   1

Professor Jamie A. Davies

What is a mammal?   2

Why do mammalian synthetic biology?   6

How is mammalian synthetic biology done?   8

What is easy, and what is hard?   11

2   Special features of mammalian systems   15

Professor Jamie A. Davies

Mammalian genes   16

Mammalian cells   22

Mammalian cells in culture   26

In vivo issues   29

The problems versus the power   30

3   Technologies for mammalian synthetic biology   33

Dr Leonard J. Nelson and Professor Alistair Elfick

We can read DNA very well   34

We can edit DNA pretty well   37

Rewriting approaches—genome editing tools   37

CRISPR/Cas9 revolution   37

Assembly of DNA parts   39

We can write DNA quite well too   39

But our capacity to author DNA remains modest   41

The next challenge is getting the modified DNA into mammalian cells   42

Getting the DNA to stay in the cell can be difficult too   42

Mammalian cellular repair machinery   45

Finally, we need to know our modified DNA is functional   45

4   Mammalian synthetic biology as a research tool   50

Professor Jamie A. Davies

Synthetic tools for analytic biology   51

Building to test ideas   56

5   Teaching mammalian cells to make new, useful things   66

Jamie Billington, Anna Mastela, and Professor Susan J. Rosser

Therapeutic proteins: the next generation of drugs   67

Cell factories for protein production   69

Mammalian cell lines   72

Using synthetic biology to improve cells for bioproduction   75

Producing therapeutic proteins in the body   76

Going agricultural: synthetic biology in farm animals   79

6   Synthetic biology, stem cells, and regenerative medicine   84

Professor Steven M. Pollard

What are stem cells?   85

Stem cell differentiation   89

Transcription factor 'master regulators' and reprogramming   89

Synthetic transcription factors   92

The road ahead   95

7   The ethics of synthetic biology   97

David Obree

Ethics overview   98

From analysis to synthesis   99

The appeal to nature   100

Potential benefits, potential harms, and uncertainty   101

Human enhancement   103

The environment   105

Responsibility and safety   106

Justice and fairness   107

Conclusion   108

Glossary   111
Bibliography   117
Index   119

ABBREVIATIONS

ATP	adenosine triphosphate
Cas9	DNA endonuclease enzyme CRISPR-associated protein 9
cGMP	cyclic guanosine monophosphate
CHO	Chinese hamster ovary
CNO	clozapine-*N*-oxide
CRISPR	clustered regularly interspaced short palindromic repeat
dNTP	deoxynucleotide
ddNTP	dideoxynucleotide
DHFR	dihydrofolate reductase
DNA	deoxyribonucleic acid
Dopamine	3,4-dihydroxyphenethylamine
DSB	double-stranded break (in DNA)
E. coli	*Escherichia coli*
EMA	European Medicines Agency
EMMA	Extensible Mammalian Modular Assembly Toolkit for the Rapid Design and Production of Diverse Expression Vectors
EP	excretory pathway
ES(C)	embryonic stem (cell)
FDA	Food and Drug Administration (USA)
GMP	good manufacturing practice
GOI	gene of interest
GPCR	G protein-coupled receptor
GPS	global positioning system
gRNA	an RNA guide used to programme Cas9 to a particular complementary location on the target (host) DNA
GS	glutathione synthetase
GTP	guanosine triphosphate
HEK	human embryonic kidney
HR	homologous recombination
HSC	haematopoietic stem cell
Indels	insertion or deletion nucleotides (at specific locations in a sequence of genomic DNA)
iPS(C)	induced pluripotent stem (cell)
IRES	internal ribosome entry site
KO	knockout
MAb	monoclonal antibody
MAPK	mitogen-activated protein kinase
mRNA	messenger RNA (see 'RNA' for that abbreviation)
MTX	methotrexate
NGS	next-generation sequencing
NHEJ	non-homologous end joining
PCR	polymerase chain reaction
PS	phosphatidylserine
PTM	post translational modification
R&D	research and development
RASSL	receptor activated solely by synthetic ligands
RNA	ribonucleic acid

RNP	ribonucleoprotein
TALEN	transcription activator-like effector nucleases
TNF	tumour necrosis factor
VEGF	vascular endothelial growth factor
ZFN	zinc-finger nuclease

1 AN INTRODUCTION TO MAMMALIAN SYNTHETIC BIOLOGY

Professor Jamie A. Davies

Learning Objectives

- Describe the features of mammals that set them apart from other animals.
- List five features of mammalian synthetic biology not found in other branches of synthetic biology as a whole.
- Give an overview of how mammalian synthetic biology is done, *in vitro* and *in vivo*.
- Make a reasoned judgement about how difficult a particular synthetic biology project would be in a mammalian system.

For most of its history, research into mammalian biology has been **analytic**. Researchers have analysed the structure, function, behaviour, development, and evolution of mammals, and have used the body of knowledge they gained to propose principles and theories about how mammals work, and what it is to be a mammal. Many analyses have focused on observing mammals entirely in their natural state, while others have involved controlled experiments on normal animals, for example, cooling a room and studying the metabolic and behavioural changes that take place to ensure that the animal remains warm. In recent years, a significant body of research has involved making single alterations to the genome of a mammal (usually a mouse), for example, deleting a gene to discover something about that gene's normal function by determining what aspects of the mammal no longer form or work properly. Though removing a gene is arguably more 'destructive' than 'synthetic', this type of work has marked the beginning of a shift away from working with only normal mammals to the construction of ones that have been deliberately altered.

Two recent developments have seen the genetic alteration of mammals move from being merely destructive interventions made to test theories to being truly constructive projects that confer on a mammal an ability it did not have before. These developments are (a) a reasonably good understanding of the molecular biology that underpins mammalian life, gleaned from many decades of painstaking analytic work; and (b) the invention of technologies for **reading, writing, and assembling deoxyribonucleic acid (DNA)** cheaply and reliably. This idea of building new genetic 'circuitry' into mammalian cells or whole mammals is called **mammalian synthetic biology**. It is in its infancy but it has already produced results of great interest both to pure science and to practical human and veterinary medicine. The field has the usual challenges inherent in all synthetic biology, and also special challenges that emerge from the fact that engineered parts have to operate in the context of a complicated multicellular organism, perhaps even a human. This fact brings with it not only technical difficulties, but also important ethical questions. These extra dimensions are the reason that this series of Oxford Primer books on synthetic biology has a separate book devoted to mammals, including, of course, ourselves.

This chapter outlines the biology of mammals, and gives an overview of the ways that synthetic biology is being applied to mammals, and why. The main purpose of the chapter is to give the reader the essential background needed to understand the rest of the book.

What is a mammal?

Mammals are a class of vertebrates (animals with backbones). The other classes of vertebrates are fish, amphibians, reptiles, and birds. The feature that makes mammals different from other types of vertebrate is that mammalian mothers feed their young on milk made in mammary glands: no other type of animal has mammary glands. Most mammals live on land, and most have a body plan consisting of a head, a neck, a trunk, four limbs, and a tail, although the proportions of these parts can vary hugely between species. Mammals are homeothermic, meaning that they maintain a steady body core temperature (typically 37°C) across a wide variety of environmental temperatures. In the cold, they use physiological mechanisms such as shivering and behavioural mechanisms such as burrowing to remain warm. In environments warmer than they are, they lose heat by panting and sweating, and by seeking shade. Most are very hairy, although some, such as humans, have only sparse hair on most parts of the body.

Mammals have the most complicated brains of any animal class, and show rich behaviours, including problem-solving. Many mammals are social and some are common companion animals for humans. Most humans make emotional connections to other mammals far more readily than they do to non-mammals: this fact, and our recognition that mammals have advanced brains of the same basic type that we have, makes the question of what we should and should not do to our fellow mammals an important one.

Internal organization

We normally describe the structure of mammalian bodies in terms of a **hierarchy of scales** (Figure 1.1). At the smallest scale, around a millionth of a millimetre across, are the molecules from which the body is made. One very important class of molecules, the **proteins**, have structures specified almost directly by genes. Proteins are responsible for running almost all of the biochemistry of the body and also for producing most of its structures. These tiny molecules, about a hundred-thousandth of a millimetre in length, are critical for life, but are not themselves alive.

Cells, typically a hundredth of a millimetre in length, are the smallest units that are considered to be alive: they contain their own collections of proteins and, collectively, these proteins run the activities of the cell itself and also perform tasks for other cells in the rest of the body. Even small adult

Fig. 1.1 The hierarchy of scales by which the mammalian body is described.

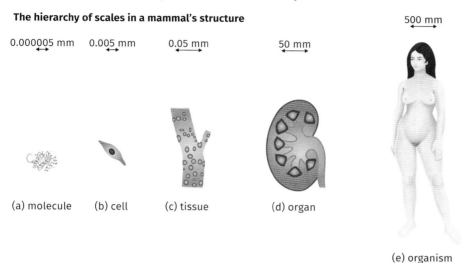

The hierarchy of scales in a mammal's structure

0.000005 mm	0.005 mm	0.05 mm	50 mm	500 mm
(a) molecule	(b) cell	(c) tissue	(d) organ	(e) organism

(a) Adapted from https://commons.wikimedia.org/wiki/File:TriosePhosphateIsomerase_Ribbon_pastel.jpg. © Dcrjsr/Wikimedia Commons/CC-BY-3.0.

mammals contain millions of cells, and these cells specialize (**differentiate**) into a few hundred distinct cell types that perform particular tasks. Some cells even specialize to perform the exact opposite task of other cells. For example, cells of the mammary gland make milk and those of the intestines digest it, but this does not matter because these particular cell types are found only in specific, separate locations of the body. This arrangement is called **compartmentalization**.

The separation of different functions in the body is made possible by cells being incorporated into tissues, which are assemblies of different cell types that together perform at least one specific function. Tissues in turn exist as constituents of organs, which are again specialized. The heart, for example, is specialized to pump blood, the lungs to oxygenate it, the intestine to digest food, the skin to provide a barrier for the body and to cool it, and so on. Typically, each organ operates semi-independently, running many of its internal processes automatically but communicating with other organs for activities that have to be coordinated between them. This communication may be chemical, for example, using hormones, or it may use the nervous system. Although the function of each organ is specialized, the same type of tissue may turn up as a constituent of many different organs, because the function of that tissue is important to all of them. Blood vessels, for example, are a constituent of almost all tissues.

 Key point

Mammalian bodies are hierarchically organized, multicellular structures in which different functions are performed in different cells and different places.

Development

Like all vertebrates, mammals develop from a single-celled zygote, produced by the union of female and male gametes (egg and sperm). The zygote divides to produce daughter cells and these in turn divide repeatedly, to produce the millions or more cells of the newborn animal. As this division takes place, cells begin to differentiate into specific types that will produce different parts of the body, in which yet new types of cells will be made. The production of these cell types follows a branching tree rather like a family tree: the first branches of this tree are shown in Figure 1.2. Indeed, the term 'stem cell' comes directly from the idea of the base of a tree of cell types. The stem cell at the base of the whole tree can give rise to everything in the body (and the placenta, etc.)—because it has the power to make anything, it is referred to as **totipotent**. Along the branches, cells become more restricted in what they can make (**pluripotent**, meaning they can make the body but not external things such as the placenta, and **multipotent**, meaning they can make more than one cell type but less than all cell types in the body). The concept of stem cells is explored further in Chapter 6. One of the earliest divisions in the tree of cell types separates the cells that can give rise to sperms or eggs (the germ line) from those that make the rest of the body (**somatic cells**). This separation means that any changes synthetic biologists make only to the somatic cells of a body will not be passed to future generations, whereas any that are made to the germ line will.

The development of mammals is broadly predictable, in the sense that the general arrangement of body parts is always the same in normal individuals,

Fig. 1.2 The branching 'tree' by which cells of the early embryo take on different identities as they proliferate. The 'tree' extends upwards through many more branch points to give rise to the approximately 250 cell types in an adult.

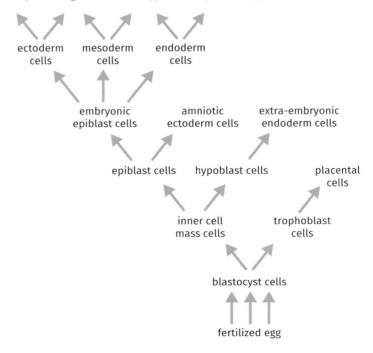

but it is not predictable down to the level of individual cells. In vertebrates in general, development involves many conversations between cells, conducted in the language of signalling molecules and hormones, and feedback. If cells in a part of a developing mammal suffer from low levels of oxygen, for example, they produce a signalling molecule called vascular endothelial growth factor (VEGF) and this attracts new blood capillaries into the area, bringing blood flow and therefore oxygen. The detailed anatomy of the capillaries is therefore not specified in advance in some genetic plan. Instead, it emerges from the exact times and places that growing cells request more capillaries for their area. The micro-anatomy of tissues therefore differs even between genetically identical individuals; this is illustrated by the fact that forensic scientists have known for many years that genetically identical twins leave different fingerprints. Genes are important to development, but they work by specifying the structures of proteins (including those such as VEGF) that conduct the cellular conversations, and the proteins that mediate the response to these conversations (e.g. making bone). They do not directly specify the structure of a large body feature such as a hand or an eye. This has important consequences for synthetic biologists, which is explored later in this chapter.

 Key point

Genes are important to development, but the relationship between genes and final structure is far from direct. It arises from cell communication and feedback events that use protein-based communication systems specified by the genes.

The development of mammals stops at a time predictable for each species, and development of new things mostly gives way to **maintenance**, although all the way through life new things are made in the two body systems that can learn from experience: the brain and the immune system. The defined time of development also means that mammals grow to a maximum size, characteristic of each species, perhaps with a difference between males and females of that species. Then growth stops, although exercise and over-eating can add muscle or fat bulk respectively, at least in domesticated environments.

Maintenance

Mammals are long-lived organisms so must renew their tissues by replacing damaged cells. Mostly, this **renewal** relies on the activity of relatively small numbers of **stem cells**, which are located in safe places and proliferate to maintain their own numbers and to produce daughter cells that will differentiate and replace lost mature cells. Most tissues have stem cells, and these stem cells can generally produce only cells of that tissue. Our understanding that tissues are renewed by stem cells has been very useful to synthetic biologists, especially those who are working in the medical field and hoping to correct a genetically abnormal tissue in a patient; they may only have to correct the genetic problem in the stem cells, and let them then replace the tissue over time with fresh, corrected tissue. This would be easier than having to correct every one of the millions of mature cells of the tissue.

Key point

Removing a sample of stem cells that maintain a tissue from the body, engineering them with a synthetic biological module, then re-introducing them into the body is a potential method for ensuring that at least some of that tissue will be replaced by cells carrying the synthetic device.

Part of body maintenance is self-defence, specifically defence against viruses, bacteria, and fungi that will infect and prey on the body if given the chance. As well as having passive barriers such as the skin, the body has an active **immune system** that patrols it and (mostly) destroys anything that should not be there. The immune system has two main components, **innate immunity** and **adaptive immunity**. The innate system, which is far older than mammals, has cells that can detect chemical 'signatures' of some bacteria, and that can also detect if a body cell is damaged. On detecting either of these things, the innate immune cells release signalling chemicals that recruit an army of cells into the area that partly close it off, destroy invaders, and clear away damaged cells. This is **inflammation**, characterized by redness, heat, swelling, and pain. The other, newer, component of the immune system is adaptive immunity; in this system, populations of cells learn from previous exposures to microorganisms to recognize them quickly and to recruit the 'thugs' of the innate system to make a very fast job of repelling the invaders. Vaccination works by teaching the adaptive immune system about human bacteria or viruses so that it knows about them even the first time it meets them for real.

In broad terms, the immune system reacts badly to anything new in the body, especially if it is associated with any injury and especially the second time it sees it. This can be a particular problem for synthetic biologists designing new things to appear in the bodies of patients they (with doctors) are trying to treat, because the body may mistake the helpful addition for an unwanted invader and destroy it.

Why do mammalian synthetic biology?

The approximately 5500 different mammalian species constitute, in numbers of species, less than four per cent of the animal kingdom, yet they dominate biological research. The main reason for this is that we ourselves are mammals, and so are most of the animals that (non-vegan) humans use as sources of food. Understanding how mammals work, and learning how to manipulate their physiology, is therefore very important to medical and agricultural research, as well as to satisfying our curiosity about our own human condition. The effort put into **biomedical research** on mammals, particularly for the most common laboratory animals such as mice, has generated large stores of knowledge such as genome sequences, and powerful tools such as methods to generate genetically modified individuals. The existence of these tools makes the mouse a particularly attractive organism for addressing questions about vertebrates as a whole, further concentrating effort into the mammalian field in a self-reinforcing cycle of research and tool development.

The drive to apply synthetic biological techniques to mammals arises mainly in response to **medical need**. Although human bodies are generally robust, self-repairing, and long-lived, they are prone to some modes of failing that are very difficult to repair with conventional medicine. These include **congenital abnormalities** in which parts of the body do not form properly, inherited **metabolic diseases** that omit some important aspect of normal biochemistry, **autoimmune diseases** in which the body attacks part of itself, and **degenerative diseases** in which parts of the body deteriorate and cannot be replaced. Some of these categories overlap: **type 1 diabetes mellitus**, for example, results from the body attacking its own insulin-producing islet cells in the pancreas. The lack of insulin alters body biochemistry and the altered biochemistry causes degenerative changes in blood vessels, eventually causing irreparable damage to organs such as kidneys.

Some of these diseases can be treated with normal medicine, and some cannot. Even where treatments exist, for example, being injected with insulin to compensate for islet loss in diabetes, they are often unpleasant and perform poorly compared to the natural system. This causes synthetic biologists to speculate about whether they can design an alternative system and place it in the body to replace what is missing. In the case of type 1 diabetes mellitus, for example, might it be possible to construct an artificial sugar-sensing and insulin-secreting system in some cell type that is not a pancreatic islet cell and will not therefore be attacked by the rogue immune system? If this could be made to work, patients could throw away their insulin syringes and their fears of eventual kidney failure, limb loss, or blindness, and live a normal life.

This, and many other examples of medical applications of synthetic biology, will be described in more detail in Chapter 6.

 Key point

One important purpose for mammalian synthetic biology is to build genetic devices that will treat chronic human diseases.

Many diseases, including some metabolic diseases and a whole range of diseases grouped under the term **cancer**, could in principle be treated by complex biological molecules such as proteins. In the case of a metabolic disease in which an important enzyme is missing, treatment might be achieved simply by producing a working version of the enzyme outside the body, for example, in a drug factory, and injecting it every so often into the patient. In the case of at least some cancers, carefully designed and engineered proteins can be constructed that cause the body's own immune system to recognize the cancer as a 'foreign' invader and kill it. Natural mammalian proteins are often folded in a complex way and equipped with **sugar side chains** that are important to their function. The folding and the addition of sugars are performed using systems inherent to mammalian cells but not present in, for example, bacteria. For this reason, trying to produce useful therapeutic proteins (such as antibodies) in bacteria often fails, and there is instead a need to make them in cultured mammalian cells. Manufacturing these proteins (**biologics**, in the language of the pharmaceutical industry) need to be done in the most efficient way, so that plenty of protein is made but not so much that manufacture drains so many resources from host cells that they fail to thrive. Organizing this often requires complex regulatory systems to be constructed, and this is the realm of synthetic biology.

Another reason for performing synthetic biology in mammals is closer to pure scientific research than to medical application. Centuries of analytical study have given us many theories about how mammals work but, while the theories are often simple and elegant, the details of the real mechanisms are usually complex and messy. One powerful way of testing whether our theories are correct is to use them to design and build simple genetic systems that ought to perform a particular function. Really building them and testing them then informs us whether our theories are correct, or whether we need to go back to the lab and ask more questions. This use of synthetic biology is explored further in Chapter 4.

How is mammalian synthetic biology done?

There are many ways of doing mammalian synthetic biology, but they can be divided for convenience into two very broad classes: engineering cells in culture (*in vitro*, from the Latin for glass) or engineering a whole living animal (*in vivo*, from the Latin for life). Both may be accompanied by computer modelling ('*in silico*', a much-used joke formulation based on the foregoing terms, that does not itself make grammatical sense in Latin).

In vitro work

Most mammalian synthetic biology begins *in vitro*, and much of it remains there. Adding new genetic modules to cells grown *in vitro* can be useful for the manufacture of proteins, for example, to be used as 'biological' drugs (see Chapter 5), for **testing theories** about how cells work (see Chapter 4), and for building devices to perform tasks such as **sensing of chemicals**.

There are two main problems with doing synthetic biology *in vitro*. One problem is to do with the way that cells incorporate new genetic material. There are many different ways to get genes into cells (see Chapter 3) but what they all have in common is that they are nothing like 100 per cent reliable. Many cells take up no genetic material at all and, of those that do take it up, many either fail to incorporate it into their chromosomes so it is eventually lost, or they incorporate it in the wrong place. Any attempt to engineer all cells in a dish will therefore result in a mixture of cells that are engineered correctly and cells that are engineered wrongly or not engineered at all. This problem is solved by **selection** of the correctly engineered cells, and growing these while the others die. Often this can be arranged to be automatic. For example, the synthetic biology modules might, in addition to performing their main function, confer on the cells an immunity to a poison that would otherwise kill them. Adding the poison to the culture, after cells have been given the synthetic module, will kill all of the cells that failed to take it into their chromosomes properly, leaving only those that did so to live on and thrive. Even then, there may be unwanted differences between these cells, so it is common to take just one of them into a new culture vessel and to allow it to multiply until it has created a whole new population—a **clone**—of identical cells. You can see a summary of these steps in Figure 1.3.

The other common problem with doing synthetic biology *in vitro* is that not all mammalian cell types can be grown in culture. Many simple types can be, and thousands of mammalian cell lines exist, together representing many types, that will grow forever if kept properly fed. Complex and specialized cells from the mature body, though, will either not grow at all in culture or they lose their specializations (**de-differentiate**) first. If it is critical that a synthetic biology module is tested or used in the context of one of these cells, the only way of

Fig. 1.3 Typical steps in engineering synthetic biological devices into mammalian cells *in vitro*: the device is introduced into a cell line, cells that have taken it up are selected, and these cells are used to become founders of clones of engineered cells.

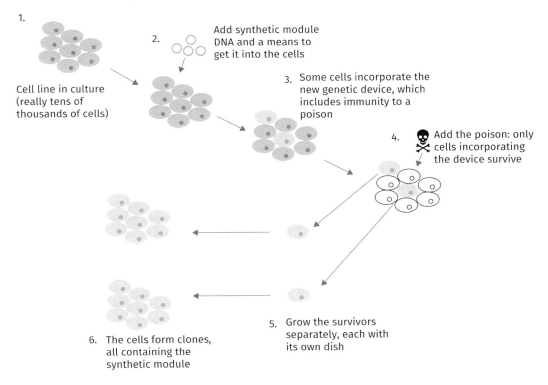

1.

Cell line in culture
(really tens of
thousands of cells)

2. Add synthetic module DNA and a means to get it into the cells

3. Some cells incorporate the new genetic device, which includes immunity to a poison

4. Add the poison: only cells incorporating the device survive

5. Grow the survivors separately, each with its own dish

6. The cells form clones, all containing the synthetic module

doing this might be to build a whole animal, every one of whose cells contains the module. To prevent it being active in irrelevant cells, it will be necessary to add to the module a genetic switch that activates it only in the correct cell type.

 Key point

One advantage of working *in vitro* is that it avoids experimentation on living animals or patients, and therefore avoids most or all ethical and legal restrictions that protect living beings.

In vivo work

Most *in vivo* synthetic biology begins with cells engineered and prepared *in vitro*, as previously described. The transition into the *in vivo* environment can be as simple as implanting a population of engineered cells into the body of a living mammal, either free to spread or **encapsulated** in some way so that oxygen, food, hormones, and other molecules can pass between the engineered cells and the host body but the cells are confined to an easily removed device. Cells carrying synthetic biological devices intended to replace missing metabolic or hormone-making capabilities, for example, to treat type 1 diabetes mellitus, can work perfectly well in such devices, and confining them reduces the risk of engineered cells causing problems by running wild. Cells engineered to fight infections or cancers need to be able to roam around the body to find their targets. So far, this type of work is mostly experimental but the researchers doing this work hope that, when the methods have been improved and shown to be safe, cells modified by synthetic biological techniques may be used to fight serious diseases in humans.

For research purposes, in particular, there is strong interest in constructing mice that have been genetically modified to carry the module in the genome of all of their cells. This is done either because there is no way to culture cells of interest outside the body, or because the point of the synthetic module is to work in a whole animal, for example, to protect it from a disease.

The most common and convenient way to create genetically modified mice is to begin *in vitro* with a culture of embryonic stem (ES) cells. These cells are obtained from very early mouse embryos—let's say a black mouse for the purposes of discussion—and cultured in a way that keeps them effectively frozen in time, always being about to carry on making an embryo, but never actually making the next step. ES cells can be engineered and cloned as with any *in vitro* cell line but, if they are placed in the early embryo of a normal mouse—let's say a white mouse—they will join in with the task of making a new animal as if they had been part of that embryo all along. The mouse that grows from the embryo is therefore a mixture (a chimaera); some of its cells come from the white-mouse cells of the embryo, and some from the black-mouse ES cells. Such mice even look like mixtures, with patches of black and of white hairs being present in their coats according to the origin of the underlying skin. If the chimaeric mouse is a male, and at least one of its testes contains sperm-making cells from the black, ES cell donor mouse, some of its sperm will carry the genes of that mouse. Some of its offspring will therefore be black and stand a good chance (typically fifty per cent, depending on how the genetic engineering was done) of carrying the synthetic biology module. If they carry it at all, it will be in all of their cells, including the cells that will produce sperm or eggs to make the next generation. See Figure 1.4 for a summary of the steps involved in making a genetically modified mouse. Embryonic stem cells are described in more detail in Chapter 6.

Fig. 1.4 The sequence of steps involved in making a genetically modified mouse from ES cells.

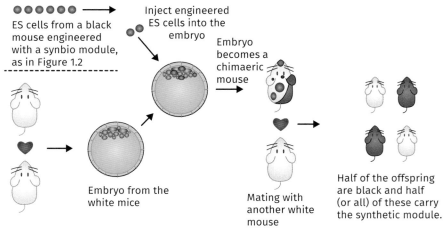

ES cells from a black mouse engineered with a synbio module, as in Figure 1.2

Inject engineered ES cells into the embryo

Embryo becomes a chimaeric mouse

Embryo from the white mice

Mating with another white mouse

Half of the offspring are black and half (or all) of these carry the synthetic module.

It is often not convenient, and may be dangerous to the animal, to have the synthetic biological module active in all cells. Genetic engineers therefore often add extra switches to their modules that will switch them on only in a specific cell type. They might also add the genetic equivalent of a 'volume control' so that the module is activated in proportion to the dosage of a harmless drug administered to the animal; that way, experimenters can take care not to have their module running so powerfully that it runs down the energy reserves of a cell and damages it. This is a useful precaution, because we do not yet understand animal physiology well enough to know what extra demands we can make of cells without upsetting their ability to do their normal jobs.

What is easy, and what is hard?

The fact that we can now read and write DNA with relative ease, place it permanently into a mammalian genome, and even make a complete animal carrying complex genetic constructions, may give a false impression about how much we can do. It is easy to be carried away with science-fiction-type ideas: why not engineer the genome to give a body gills that can breathe under water? Or how about having extra eyes in the back of the head for genuine all-round vision? Or how about including a spare heart that can take over at a moment's notice if the main one has a heart attack?

This type of thing, if it ever happens (and there are ethical as well as technical issues to consider), is a long way off. See Chapter 7 for further discussion of ethical issues. The problem is that although the structure and functions of the body are in some way specified by the genes, this specification is very indirect and a world away from the way that the machines we make are specified by wiring diagrams and engineers' drawings: that cliché 'genetic blueprint' is a very misleading metaphor. Genes do specify most of the properties of individual proteins and ribonucleic acid (RNA) molecules in a direct way. That is all they specify. Everything else the body makes or does **emerges** from the complex interactions of these RNAs and proteins with one another, with the environment, and with the genes. The shape of your nose and the firmness of your biceps do owe something to your genes but there is no 'shape of nose' gene or 'firmness of biceps' gene to edit. Rather, the interactions of thousands of proteins, made by thousands of genes, in millions of cells, determined the shape of that nose. Even though a scientist could read your genome easily now, our current state of knowledge means she would have very little chance of being able to deduce, from that genome, anything about your face beyond general information about skin pigmentation and eye colour. In the case of the firmness of your biceps, whether you live a quiet, bookish life or work out in the gym every day will have a far greater effect than would the genes you inherited, so the **environment** and **behaviour** are important too. Given that we do not understand the relationship between our genes and even the normal tissue- and organ-level structures of our bodies, building a synthetic biological module to add a new designer organ or body appendage to either a mouse or a human is not yet feasible. If it ever will be so, it will depend on a lot more knowledge coming from basic, analytic biology.

If we cannot use mammalian synthetic biology to make outlandish designer bodies, what can we do? We do understand well the relationship between genes and proteins, and we can certainly use **designer genes** to make **designer proteins**. This is useful for two related reasons. The first is that proteins, particularly that class of catalytic proteins called enzymes, run the chemical reactions of the body's metabolism. They also run most types of communication between cells, either directly or by making non-protein hormones. Therefore, we can make genetic modules that add new metabolic pathways to mammalian cells, for example, to make vitamins that can be made by other organisms but which the animal in question must obtain from its food (or suffer a deficiency disease if it cannot do so). We can also make new proteins designed to kill viruses or bacteria, or to detect disease and to report it.

Many diseases of humans arise because of defects in metabolism or hormonal signalling. Signalling mechanisms are discussed in more detail in Chapter 4. These may occur because of a genetic defect present from conception; porphyria, the disease responsible for 'the madness of King George', is an example that results from a mutation in the gene encoding a metabolic enzyme. Alternatively, an initially healthy body may be damaged so that the tissue that should make an enzyme or hormone is no longer able to. Type 1 diabetes mellitus results from the immune system of the body mistakenly destroying pancreatic islet cells as if they were foreign invaders, so that the insulin normally made by these cells can no longer be made. Engineering synthetic biological modules that make insulin, at the right levels for the sugar in the blood (this is important—too little insulin or too much are both dangerous), is the type of work that is challenging but is in principle within the reach of current synthetic biology techniques (see Case study 1.1). Examples of this are discussed further in Chapter 5.

 Key point

Synthetic biological projects designed to add metabolic or signalling functions to a mammal are broadly feasible with current technology. Those aiming to perform major redesigns of the body are broadly not feasible.

Case study 1.1
A synthetic morphology device to combat diabetes

Mingqi Xie and colleagues, in Basel, constructed a synthetic biology system that could sense the levels of glucose in body fluids and secrete insulin when they became high. They introduced this into the body of mice that had type 1 diabetes mellitus, and showed that their high blood sugar fell to almost normal levels within three days. This is a demonstration of the type of approach that might, one day, be used to control human disease.

Chapter summary

- Mammals are complex, hierarchically organized, multicellular organisms in which different functions are separated into different body compartments.
- Because we humans are mammals, a major reason for performing mammalian synthetic biology is the development of new medicines and genetic devices to treat human disease.
- Mammalian tissues are mostly built and maintained by stem cells.
- Mammalian proteins often carry sugar side chains that cannot be added by other, more commonly engineered cell types such as bacteria; this is one reason why large-scale production of mammalian-type proteins often depends on engineering genuine mammalian cells to make the proteins.
- *In vitro* mammalian synthetic biology usually involves introducing a synthetic genetic device into cultured cell lines, selecting those cells that incorporated it properly, and growing clones of those cells.
- *In vivo* mammalian synthetic biology may involve implanting cells engineered *in vitro* into a host animal, perhaps encapsulated.
- *In vivo* mammalian synthetic biology may alternatively involve engineering a synthetic module into ES cells *in vitro*, incorporating them into an early mouse embryo, allowing that embryo to develop into a chimaeric adult, breeding from that adult, and identifying pups that carry the synthetic module.
- Using synthetic biology to engineer novel metabolism or cell signalling is feasible with current technology.
- Using synthetic biology to make radical, functional changes to the mammalian body plan is largely beyond current science and technology.

 Further reading

Black JB, Perez-Pinera P, Gersbach CA (2017). Mammalian synthetic biology: engineering biological systems. *Ann Rev Biomedical Engineering* 19, 249–77.
This is a detailed review of the general potential of mammalian synthetic biology, written at a moderately technical level.

Leonard JN (2014). The rise of mammals. *ACS Synth Biol* 3, 878–9.
A short, readable overview.

Xie M, Ye H, Wang H, Charpin-El Hamri G, Lormeau C, Saxena P, et al. (2016). β-cell-mimetic designer cells provide closed-loop glycemic control. *Science* 354, 1296–301.
This describes the results in the case study, but be warned that it is a difficult read intended for professional scientists.

 Discussion questions

1.1 What are the advantages and disadvantages of working *in vitro* or *in vivo*?
1.2 A synthetic biologist wants to engineer a synthetic genetic module into the genome of a mouse, using the standard method of making a chimaeric mouse from ES cells, depicted in Figure 1.4. Would there be any advantage in him engineering it deliberately into the same chromosome that controls coat colour?

1.3 Which of the following projects are probably within the grasp of current knowledge and technology, and which would not be (ignore ethical considerations): (a) making a device that senses uric acid build-up in the metabolic disease gout, and secretes an enzyme to destroy the excess acid; (b) making a six-legged horse for more interesting dressage events; (c) making a device to give humans the metabolic capability to make their own vitamin C; (d) making a device that produces glucocorticoid hormones, in response to low blood glucose levels, for patients who cannot make these hormones for themselves because their adrenal glands are damaged; (e) adding whiskers to humans so that they can avoid bumping into things in the dark?

SPECIAL FEATURES OF MAMMALIAN SYSTEMS

2

Professor Jamie A. Davies

Learning Objectives

- List differences and similarities between bacteria, yeasts, and mammals, in features relevant to the design of synthetic biological devices and projects.
- Describe the typical structure of a mammalian gene, including its controlling elements.
- Explain how transcription of a typical mammalian gene is activated.
- Explain how mammalian gene expression can be controlled by chromatin modification, and the significance of this to synthetic biology.
- Give an outline sketch of how mammalian gene expression can be specific to particular cell types.
- List or sketch the major compartments in mammalian cells, and label each with at least one function.
- Describe the main barriers to the spread of biomolecules within the mammalian body.
- Describe in outline how the mammalian adaptive immune system works and how its presence can be a limitation to synthetic biology.
- List the special problems of maintaining mammalian cells in culture.
- Explain the specific safety considerations critical to working with human cells.

The first synthetic biological devices that most students build are designed to operate in simple organisms such as bacteria. There are many reasons for this: bacteria are cheap, simple to look after, and easy to engineer; they grow and multiply quickly enough for a small project to be done in a week or two; and work on bacteria raises few ethical concerns. Familiarity only with bacteria does not, however, prepare a synthetic biologist for mammalian systems because mammals have many features bacteria do not have. Some of these features are common to all **eukaryotic** cells, even very simple ones such as yeast; some are common to all animals; and some are unique to 'higher' animals such as mammals. We consider all of these types of differences in this chapter, noting in each case the range of organisms that share a special feature. Figure 2.1 shows a summary of these differences.

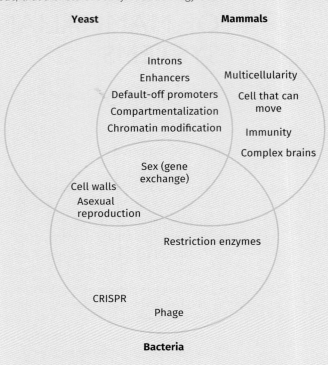

Fig. 2.1 Selected features of bacteria (prokaryotic), yeasts, and mammals (both eukaryotic) that are relevant to synthetic biology and are discussed in this chapter.

Mammalian genes

The genes of mammals are similar to those of other eukaryotes (organisms whose cells contain nuclei), but differ in several important respects from genes in eubacteria such as the much-used laboratory bacterium, *Escherichia coli*. Some basic things are the same in all organisms: genes specify the structure of proteins using a **triplet** of three DNA bases to specify each amino acid. The relationship between these triplets and the amino acids for which they code is the 'genetic code', and it is the same for all natural organisms. Also, in all organisms, the information in the DNA of a protein-coding gene is **transcribed** first into RNA, which is then **translated** into protein by ribosomes. Some genes do not code for proteins: their RNA products are directly active in the cell, for example, as constituents of ribosomes. Beyond those basic features, though, there are differences.

Mammalian cells have two copies (or near-copies) of most genes

Bacteria have one chromosome and, usually, one copy of each gene. Mammalian cells have a number of different chromosomes, arranged in pairs: women have twenty-three pairs, for example (men have twenty-two pairs and one non-pair, the X and Y). When cells proliferate, either in a culture dish or during growth of the animal,

Fig. 2.2 Chromosomes in mammalian reproduction: chromosomes of the same type (e.g. the two copies of chromosome 1) are depicted in the same colour, and for simplicity only two pairs are shown. Top panel—mammalian cells reproduce asexually, passing on copies of all chromosomes to their daughters. Bottom panel—mammalian organisms reproduce sexually, placing just one of each chromosome type in a gamete so the next generation inherits one of each type from its mother and one from its father.

(a) Asexual reproduction of cells

Daughter cells

(b) Sexual reproduction of organisms

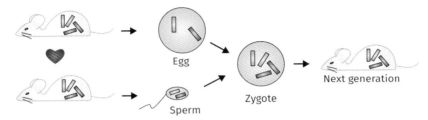

chromosomes are copied exactly (neglecting rare errors) and daughter cells are genetically identical to their mother cell (Figure 2.2a). Engineering a synthetic construct into any of the chromosomes will therefore ensure that it is passed down cell generations in culture.

When whole mammals reproduce, they do so sexually. Each parent produces gametes (sperms or eggs), each of which contains only one of each type of chromosome rather than a pair. This single representative of each type is a hybrid of the two original representatives, thanks to the process of 'recombination' involved in making sex cells. In the new child to be born of the union of this sperm and egg, one of each pair of chromosomes comes from the animal's mother, and the other from the father (Figure 2.2b). They will be similar but not identical because most genes have different versions (alleles) in the population and different members of a chromosome pair may contain different alleles of any particular gene. Sometimes, the alleles may give different instructions. The bickering of parents is therefore not merely a household matter: the genetic voices of a mother and father carry out their ceaseless argument in every one of their child's cells—*blue eyes!, no, brown!, blood type O!, no, blood type A!*, and so on. Only when that child grows up and makes their own sperm or egg cells do these parental messages finally combine to present a united front, in the grandchild, against the instructions that will come from the grandchild's other parent.

This process of random combination of mother-derived and father-derived chromosomes, for each pair, and also swapping parts between them, means that a synthetic biological device incorporated into one chromosome will be passed, on average, to only half of an animal's offspring. If the device must be passed on to all offspring, it must be present in both chromosomes of a pair, and it must be in both chromosomes of both founding parents if it is to be passed on to all of their progeny in all future generations.

While the presence of paired chromosomes in mammalian cells adds a complication, it can also be used to make engineering easier. It is possible, for example, for different laboratories to engineer different parts of a complex synthetic biological system on different chromosomes of mice, and then for them to breed the mice to bring the different parts together in (some of) their offspring.

Interrupted genes: introns, exons, and splicing

The protein-coding part of a typical bacterial gene has a simple structure and runs uninterrupted from beginning to end. There is no need for any editing of the messenger RNA (mRNA) between its transcription and translation, and translation of bacterial mRNAs often begins at one end of the mRNA before transcription has been completed at the other. Most mammalian genes, on the other hand, have a very different structure.

> **Key point**
>
> Mammalian genes have a more complicated structure than those of bacteria.

Their protein-coding regions, exons, are interrupted by several non-coding regions, introns. The RNA that is transcribed from mammalian genes contains exon and intron regions just as the gene did, and it has to undergo **RNA processing** reactions during which the introns are spliced out to leave just the exons, joined together seamlessly (Figure 2.3a). This splicing is performed by **splicing complexes (spliceosomes)**, which recognize special **splice-donor**

Fig. 2.3 The splicing of introns out of RNA. Panel (a) shows constitutive (no-choice) splicing, in which the introns, shown red, are spliced out of the RNA to leave just the blue exons to produce a continuous protein-coding sequence. Panel (b) shows alternative splicing, in this case a choice about whether to include exon 2.

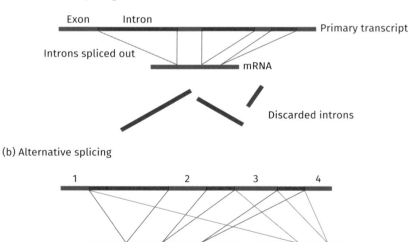

and splice-acceptor sites in the RNA that mark the ends of the introns. Being part of the introns, the splice-donor and splice-acceptor sites are removed by the splicing so do not interfere with the protein-coding sequence of the exons and, once the introns have been removed, a ribosome can read the mature mRNA as if it had been continuous all along. There is no risk of translation beginning before introns are spliced out, because transcription and splicing take place in one part of the cell, the nucleus, while translation takes place in a different part, the cytoplasm (see 'Mammalian cells' later in this chapter).

The presence of introns in mammalian genes means that they are often much longer than they would need to be. Human genes have an average of just less than ten introns in each gene and there is often far more intron than exon in the gene as a whole. To take an extreme example, the human dystrophin gene is over two million bases long because ninety-nine per cent of it is made of introns.

Given the energetic cost of 'pointlessly' transcribing introns into RNA, there must be some good reason for mammals to use introns, and at least two have been discovered. One reason is to control the relative **timing** of protein production. A typical intron-free gene takes just a few minutes to transcribe; the dystrophin gene mentioned previously takes sixteen hours. These differences can be useful. Consider three genes: 'A' with many long introns, 'B' with a few short ones, and 'C' with none. If transcription of all three genes is activated at the same time, transcription of C will be complete in a few minutes, of B in some tens of minutes, and of A in hours. For a biological process that depends on events taking place in a set sequence, for example, embryonic development, this can be very important. A 2011 experiment removed introns from the gene *Hes7*, which is important in mouse development. The result was that *Hes7* protein was produced earlier than usual and the development of the embryo was, as a consequence, very abnormal. The presence of the long intron is therefore important to the function of this gene.

Another reason for mammalian genes to have exons is that it allows several different proteins to be produced from one gene. It turns out that while many splicing sites are constitutive (they are always spliced and joined), some involve a choice of which donor site gets spliced to which acceptor site (Figure 2.3b). Choosing not the next site but the next-but-one will result in an exon being missed out altogether and a different version of the protein being produced. This mechanism for production of different versions of a protein is called alternative splicing. In really complex cases, many different versions of the protein can be produced: alternative splicing allows the rat fibronectin gene, for example, to produce at least twelve different versions of the fibronectin protein, each with distinct functional properties. The choice of which combination of splice donor and acceptor are spliced depends on the presence of particular RNAs and proteins in spliceosomes of the cell: different cell types have different combinations. This is an example of epigenetic control: 'epigenetic' is a term coined by the geneticist Conrad Waddington, and is a catch-all term for any influence that works above ('epi-') the level of the gene, rather than being the sequence of the gene itself. The set of splicing factors present in a cell is one example of an epigenetic influence, and more are discussed later in this chapter.

Genes do not have to have introns or splicing to work in mammalian cells: about eight per cent of mammalian genes do not, and synthetic genes constructed purely from exon sequences of natural mammalian genes work perfectly well.

Promoters and enhancers

 Key point

Mammalian genes are controlled by promoters and a number of distant enhancers.

Bacteria, yeast, and mammals all activate gene expression from promoters just upstream of the protein-coding region of a gene, but the action and structure of these promoters is different. Bacteria often use promoters that are constitutively active: unless the cell contains specific inhibitory proteins to stop them doing so, a promoter will bind gene-transcribing complexes and set them to work on transcribing the gene.

The promoters of eukaryotic cells, including mammals, are usually not constitutively active. They contain fairly standard binding sites for transcription complexes (e.g. the 'TATA box') close to the transcription start site, and also sequences that bind specific additional transcription factors. In general, active transcription complexes are recruited and set to work on the gene only if these additional transcription factors are present. There are many different types of transcription factor, and each will act as an 'on' switch only for genes whose promoters will bind them. Thus different transcription factors will switch on different genes, or different sets of genes.

In general, the promoters of mammalian genes activate transcription relatively weakly. Their action can be strengthened by additional transcriptional activators that bind to further elements of DNA, enhancers. Enhancers are often situated far (hundreds of bases) from the promotors they assist, and may be either upstream or downstream. Some, for example, are sited in an intron of the gene to which they apply. The linear way in which drawings such as Figure 2.2 depict DNA can be misleading: in reality, intervening DNA loops out to allow proteins bound to enhancers to interact with those bound to promoters, as shown in Figure 2.4.

Fig. 2.4 Transcription of mammalian genes is controlled by both promotors, just upstream from the transcription start site, and enhancers, which can be a long distance either upstream or downstream. Proteins that bind promoters and enhancers interact because the intervening DNA loops out.

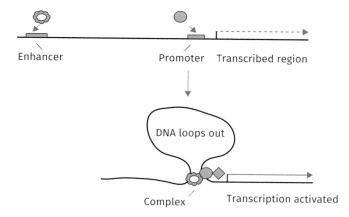

Mammalian genes (and those of other multicellular eukaryotes) often have a range of enhancers that bind to different **transcription activators** or t**ranscription inhibitors**. This means that the activity of the gene responds to the combination of the factors that are present in the cell. The effect of different activators can be additive, and the effect of an inhibitor can negate the influence of activators. Their presence in a cell is another example of an epigenetic influence. The actions of multiple transcription factors on a gene are often summarized in the language of Boolean algebra—'a gene is on if transcription factor A is present *or* transcription factor B is present', for example, or 'a gene is active if inhibitor I is *not* present'. See Figure 2.5. This way of describing things is a useful shorthand, but it should not be taken too far: transcriptional activity is an **analogue** process, which can have a full range of activities between 'off' and 'on' and modelling it as digital without good reasons for doing so can lead to poor performance of synthetic biological systems. Transcription factors are considered in more detail in Chapter 6.

Some promoters, particularly ones in viruses that prey on mammalian cells, are strongly active in most cell types. These are widely used in synthetic biology because they are simple. These, and natural promoters, can be combined with DNA sequences that bind repressors of gene expression in bacteria. This allows these repressors, if made in mammalian cells, to repress these mammalian promoters too.

Chromatin modifications

The DNA of mammalian cells exists as a set of distinct pieces, each of which is a constituent of a chromosome. Humans, for example, have forty-six chromosomes, twenty-three from each parent. In the chromosome, the DNA is coiled

Fig. 2.5 Mammalian gene expression is controlled by factors binding to enhancers acting in combination. Panel (a) depicts a gene with a promoter and two enhancers (there are often more). Panel (b) shows the way that synthetic biologists often express combinatorial actin in terms of Boolean logic gates, but panel (c) is a reminder that transcription activity is an analogue, continuously variable measure and not a digital, all-or-none one, and that between concentrations that are too low to be active or high enough to be saturating, different factors are combined by analogue and not digital computation.

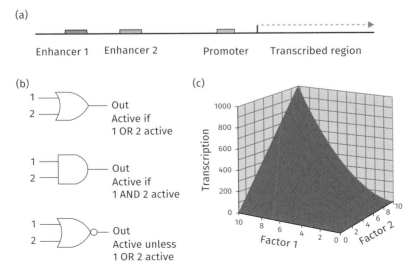

around blocks of protein called **nucleosomes**, and the coiled structure is itself coiled up and linked with other proteins to make the structure of the chromosome. This coiling up is a necessity—if all of the DNA in a human cell were laid out as a single line, it would be 2 m long (the height of a very tall man), yet being coiled up it can fit into a cell only 0.00001 m across. Coiling of DNA brings challenges to the cell, because the more compactly the DNA is wound up around proteins, the more difficult it is for transcription factors to find and bind it, and the more difficult will be the process of transcription; this is partly because of access and partly because of the need to avoid making a hopeless tangle of the new RNA and the DNA from which it was transcribed. The degree of coiling and compression—**condensation** in the language of chromosome science—is therefore different in different parts of the chromosome. Those parts that contain no active genes are generally tightly coiled (**heterochromatin**), while those with active genes are more loosely coiled (**euchromatin**). But since tight coiling inhibits gene activation, there is a need to uncoil regions of DNA to make their genes accessible to the influences of transcription factors. Hence the regulation of condensation of chromatin is another very important level of epigenetic control of gene expression.

Transitions between euchromatin and heterochromatin are mediated mainly by enzymatic modifications of the proteins around which DNA is wrapped. Addition of methyl, acetyl, and phosphate groups to these proteins changes the way that they pack. Unfortunately, there is no simple correlation between how much of this modification exists at a site and its activity, because methylation at some sites of some proteins drives chromatin into the eukaryotic, active state, but methylation at other sites has the opposite effect. The details are still being researched.

Mammalian cells

Anatomically speaking, the inside of a bacterial cell is relatively simple. True, some functions are located in specific places, where large assemblies of proteins exist to carry them out, but there are no internal barriers to the free passage of molecules. The DNA of the genome, the ribosomes that make protein, and the enzymes that run metabolism all share a common **cytoplasm**. As long as they simply want their devices to work inside a bacterial cell, synthetic biologists seldom have to worry about whether the proteins made from their genes will get to the right place, because they can and will get everywhere. Outside a bacterial cell is another story: bacteria are surrounded by **cell walls**, of varying complexity depending on the species, and targeting proteins to a specific layer of the cell wall, or arranging for them to pass through it, may take a lot of careful planning.

Mammalian cells are much larger than bacterial cells—of the order of 1000 to 10,000 times larger in terms of volume, depending on the cells involved. Like the cells of other eukaryotes, they also have a much more complicated internal anatomy. Unlike bacterial cells, they contain many membrane-bound organelles, inside which specific functions take place, separated from one another and from the bulk cytoplasm. Some are depicted in Figure 2.6. Destruction and recycling of proteins takes place in **lysosomes**: for example, replication and transcription of DNA takes place in the **nucleus**; adenosine triphosphate (ATP) generation takes place in **mitochondria**; translation of mRNA into cytoplasmic proteins takes place in the bulk cytoplasm, but translation of most proteins destined for

Fig. 2.6 A simplified diagram of the internal structure of a generalized mammalian cell. The cell is divided into compartments—organelles—each with its own function. Structures involved in reading a gene to produce a mature protein are shown in green, those involved in destroying proteins are shown in red, and structures involved in turning food into usable energy are in orange.

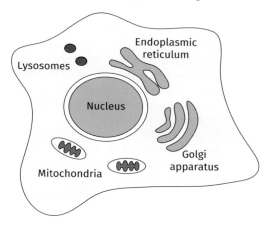

the membrane or for secretion takes place on the **endoplasmic reticulum**. Proteins bound for export then pass through the **Golgi apparatus** to be modified, for example, by the addition of sugar side chains. Most molecules in one compartment (organelle/cytoplasm) are trapped there by the membrane and cannot enter other compartments freely. Newly made proteins that are meant to go to a compartment are targeted there by an elaborate system of 'post-code' amino acid sequences that are read by cellular machinery that regulates traffic between compartments. Depending on the project, synthetic biologists may have to include such 'post-code' sequences in their proteins, and take care that everything that is intended to interact is in the same compartment.

 Key point

The design of synthetic biological systems for use in mammals must consider whether the proteins produced from them will be in the correct compartment.

The boundary of mammalian cells is much simpler than that of bacteria, yeast, or plants: there is no cell wall, just a thin and mechanically weak plasma membrane. This makes the production of exported proteins easier, but the weakness and simplicity of the membrane does mean that mammalian cells are fragile and much more difficult to grow than are bacteria and yeast.

As was mentioned in Chapter 1, mature mammalian bodies contain hundreds of different types of cell, each specialized for a particular range of physiological functions. The specialization is achieved in part by each cell type expressing a different subset of genes. Some genes are active in all cells because they serve functions that are needed by all cells; typically, they are involved in the basic metabolic reactions of life and in running the machinery common to all cells. They are called housekeeping genes. Other genes are active only in some specialized cell types, perhaps in only one. As a general rule, all cells of the body contain all genes and they achieve their different combinations of gene expression by switching some genes on and others off, rather than by eliminating genes they do not need (see Scientific approach 2.1). This switching is done through epigenetics, specifically through chromatin modification driven by specific regulatory proteins, and through the set of transcription factors made in the cell. The transcription factors and chromatin modifiers are, of course, themselves proteins and are therefore made by genes. For a cell type to be stable, it is therefore necessary for it to have a set of transcription factors and chromatin modifiers which ensure their own continued production, as well as the production of other proteins which that cell needs. This illustrates an essential feature of the interplay of genetics and epigenetics: they form a cycle of **positive feedback** to keep gene expression stable. See Figure 2.7 for a very simplified depiction of this control structure.

Mammalian cells are very sensitive to their environment which will, of course, generally be that of the inside of the body, created by the influences of neighbouring cells and, to a lesser extent, those further away. Cells make signalling proteins for short- or long-range communication, and also bear **signalling molecules** on their surfaces so that signalling from these happens only through cell-to-cell contact. As well as making one set of signals, cells bear **receptors** that make them sensitive to (usually) different signals made by other cells. The receptors are typically transmembrane proteins, which allow them to transmit news of the binding of a signal to the portion of the receptor outside the cell to the environment inside the cell. The presence of signals typically acts through receptors to activate transcription factors and, through them, gene expression. The simple diagram in Figure 2.7 is therefore made much more complex because the stability of a cell's state of gene expression depends not only on positive feedback within itself, but also on the presence of the right range of signals from the local environment. The ability of cell A to produce the right signals to maintain cell B may depend on cell B producing signals to maintain cell A, and so on. See Figure 2.8 for a simplified depiction of this.

Fig. 2.7 A very simplified depiction of the way that a set of genes can maintain its own expression and also the expression of genes that produce useful products such as enzymes, structural proteins, and signalling proteins. These are shown in blue. Transcription activators are shown in green and a transcriptional inhibitor in orange. These form networks of feedback that maintain their own expression. This diagram, unlike Figure 2.8, ignores influences from outside the cell.

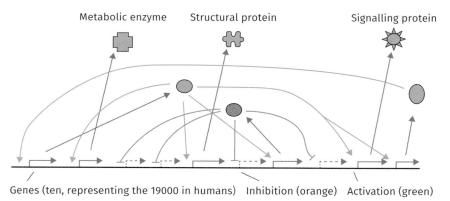

Metabolic enzyme   Structural protein   Signalling protein

Genes (ten, representing the 19000 in humans)   Inhibition (orange)   Activation (green)

Fig. 2.8 Within mammalian bodies, patterns of gene expression are maintained by integrated actions of inner feedback loops and also by the gene-controlling actions of signals that pass between them.

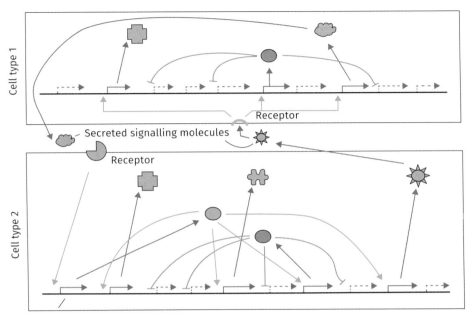

If the signals change, gene expression in the cells may change. This effect is used in physiology, and especially in development which is a process of change. It also has consequences for researchers wishing to maintain a cell type in culture outside the body (see 'Mammalian cells in culture').

Synthetic biologists may wish to have their device active in a particular cell type, for example, in milk-producing cells of the mammary gland for a project designed to produce a valuable medical protein in milk. To do this, they need to use the findings of analytic biology to understand which transcription factors are present in the cell of interest, and arrange their device to be activated by one or more of these transcription factors. They also need to ensure that their device does not have unintended consequences on the transcription of other genes, or they may find that they eliminate the very cell type they wish to work with.

Scientific approach 2.1

How do we know that specialized cells still contain the complete genome? The pioneering work of John Gurdon, in 1962, began a series of experiments that have proved this point. The experiments work by taking a nucleus from a mature, differentiated cell and using it to replace the nucleus of an egg, then showing that the egg will develop into a complete embryo and, eventually, a complete animal with all of its specialized cells. This proves that the nucleus of the specialized cell must contain the genes required by different types of specialized cell too. Gurdon's work was done in frogs; the first mammal created from the nucleus of a differentiated cell was Dolly the sheep, created by Ian Wilmut's team in Edinburgh.

Unlike the cells of plants, animal cells can move. Some types of cell, such as a skin cell, move very little in their lives while others, such as **neutrophils** of the immune system, move a great deal and patrol the body. This can be an important consideration for experimenters who alter specific cells either in a culture dish or in the body, because a cell will not necessarily remain where it was. It is therefore common for experiments to include the addition of markers such as fluorescent proteins to cells, as well as the addition of the main synthetic biological module, so that the treated cells can be identified even when they have moved.

Mammalian cells in culture

 Key point

Much mammalian synthetic biology takes place in cultured cells only and is never intended for use in an intact animal. Even work intended for use in animals often begins in culture, so techniques for culturing mammalian cells are very important to the field.

Unlike bacteria and yeast, which exist in nature as single-celled organisms that have to survive in a changeable and unpredictable external environment, most mammalian cells have evolved to function entirely within the body. The exceptions, such as the cells that line your mouth, are mostly within the body but with a small part exposed to a well-controlled mucus layer that comes between them and the real outside world. Given that they face predictable surroundings with temperature, salinity, and pH regulated closely by the physiology of the mammal as a whole, mammalian cells can trade toughness for flexibility. They have no cell wall to resist osmotic swelling and only modest systems for dealing with acidity, alkalinity, cold, or the sheer stresses of swirling liquids. Mammalian cells are, in a word, fragile.

Mammalian cell culture requires special media and gases

Bacteria and yeasts evolved to live freely in damp, nutrient-laden environments, either in the general environment or, in the case of *E. coli*, in the gut of a mammal. These environments can be more or less salty, rich in nutrients, or poor in them, with a range of different possible nutrients available to which the yeasts and bacteria need to adapt. They can be cultured simply by placing them in a shaker full of simple nutrient broth. Mammalian cells have never evolved to fend for themselves in the same way: they evolved to live in body fluids that are very complex mixtures of ions, small organic molecules, vitamins, and hormones such as insulin. **Culture media** for mammalian cells must therefore mimic these fluids very closely, or the cells will die. Typically, media are so complex that companies have been established specifically to produce them very accurately in large, quality-controlled batches, and laboratories buy these instead of trying to make media from hundreds of components themselves.

 The pH of culture medium must be controlled carefully, particularly against the tendency of cellular energy metabolism to produce acid by-products. Unfortunately, few mammalian cells tolerate synthetic pH buffers. The main pH buffer of the body is an equilibrium reaction between bicarbonate, carbonic acid, and carbon dioxide, with the enzyme carbonic anhydrase mediating rapid two-way conversion between carbonic acid and carbon dioxide plus water.

To mimic this natural system, sodium bicarbonate is a constituent of the culture medium but, in order to maintain the natural pH of the body, the gas above the medium has to contain **five per cent carbon dioxide**, a concentration similar to that in tissues and about a hundred times that in air. This requirement means that mammalian cells need complicated incubators that maintain not only temperature but also carbon dioxide concentration, taking the gas from cylinders of the same type that automobile workshops use for metal inert gas welding.

No cell is an island

Mammalian cells are equipped with receptors to detect a large range of signals coming from their environment and unless they receive a minimum set of signals correct for that precise cell type they will die in a process of cell suicide. This suicide mechanism is called anoikis, from the Greek for homelessness, and it probably evolved as a defence against body function being harmed by cells that have simply got lost during development. It is a sub-class of a more general suicide mechanism called apoptosis. Signals that promote survival are of three main types: those that are present in body fluids, those given by contact with other cells, and those given by contact with the **extracellular matrix** (the complex 'packing material' in connective tissue, which is also present as a thin layer underneath cell sheets).

Signals given in body fluids are generally **hormones** and **growth factors**. For some cells, the requirements have been identified and can be satisfied by giving a defined mix of these molecules in the medium. For many, the factors have not been identified and poorly defined additives, such as the commonly used **fetal bovine serum**, are used instead. Signals from the extracellular matrix are given by specific proteins of the matrix interacting with receptors on cells. Generally, the signalling only works properly if it comes from a surface on which the cells are growing, rather than from molecules in free solution. Most mammalian cells therefore need to grow on a solid surface such as the bottom of a glass or plastic dish, to which the correct proteins have been added. In some cases, adding the correct proteins is simple because cells make them for themselves and glass and tissue-culture grade plastics bind these proteins well. In other cases, elaborate coating protocols have to be followed to make the culture vessel ready for the cells. The need for most cells to grow on a surface means that they grow two-dimensionally as sheets, and fewer cells can be grown in a given container than if they could grow in bulk suspension.

Some cell types do tolerate growth in suspension. These include blood cells, whose natural tissue is a liquid, and also cancer cells that have been mutated so far from their natural state that they have lost the need for survival signals from a surface. It is the loss of this need that enables some cancer cells to spread in the body, and such cells are usually the most dangerous to a patient with the disease.

The quest for immortality

Mammalian cells are grown in a cycle of incubations and **passages**. During the incubation phases, which usually last forty-eight hours, a population of cells that begins by covering about one-quarter of a culture flask's surface will grow to cover it all. During passage, the cells are removed from the surface, usually using protein-digesting enzymes and small organic molecules that chelate ('mop up') divalent cations such as calcium that are needed for cells to stick to a surface. This allows the cells to be brought temporarily into suspension, when they are diluted typically by about one in four, and either divided into four more culture vessels for a further incubation, or one-quarter is put into a culture vessel and the other three-quarters are used for an experiment.

It was observed many years ago by Hayflick that cells obtained fresh from an animal can be grown for only a finite number of passages before they die. This Hayflick limit has nothing to do with the skill of the culture technician, but it is connected with the age of the animal, cells from young animals being able to grow for more passages than cells from older ones. These observations gave rise to the hypothesis that cells can divide only a finite number of times before they run out of something important. It turns out that they do exactly this. Each time a normal cell replicates its DNA, the end-caps (**telomeres**) of its chromosomes become slightly shorter, and when they are too short the cell can divide no more. Most cells have no way of rebuilding their telomeres, but cells that will give rise to sperms and eggs can rebuild them using an enzyme called **telomerase**, thus resetting the system. Some stem cells also have this property, and can be kept in culture forever. There may also be other reasons, not connected with telomeres, that cells have a limited life: this is a topic of intense ongoing research.

There are three ways round the Hayflick limit. One is to accept it, to engineer synthetic biological systems into cells from young animals, and to plan for having to do it all over again with a fresh batch of cells in a few years' time. Another is to use a **cell line**. When normal cells are passaged repeatedly, mutations sometimes arise in them that grow quickly, outcompeting the original cells in the culture, and in many cases these are immortal in the sense that they do not show any limit to the number of times they can be passaged. Usually these cells have lost some properties of the originals, and tend to behave as immature types (much as cancer cells do). Depending on the work to be done, this may not be a problem. There are large banks of cell lines maintained across the world, and they are easy to obtain and usually easy to look after. The third way around the Hayflick limit is to engineer genes into cultured cells that ensure their continued division and maintenance of telomeres. Such genes, collectively known as '**immortalizing genes**', have been obtained from various viruses, some but not all being viruses associated with cancers. Some variants of these genes exist that code for proteins active at 33°C but inactive at 37°C. In principle, cells can be maintained as an immortal population at 33°C and then warmed to 37°C to be normal again. This sometimes works, but often the cells have forgotten how to be normal. There is even a mouse, 'immortomouse', genetically engineered to have this temperature-dependent system in all of its cells. Mouse bodies run at the usual mammalian temperature of 37°C, so the system is inactive, but if cells are removed from a mouse and cultured at 33°C, they should grow without limit. This does not always work in practice, but it works enough times for the mouse to be useful.

Safety considerations

Because we are mammals, working with mammalian cells can bring with it some safety considerations that are absent when working with bacteria or yeast, especially when human cells are involved. The main danger is that a virus that can infect human beings can in principle grow in a culture of human cells, where it will not be held in check by any immune system because there is no such system in culture. Cells isolated directly from human tissue therefore need to be checked carefully for possible viruses, especially for difficult-to-treat viruses such as hepatitis and HIV and, even when clean cells are used, care needs to be taken not to contaminate them with a virus such as the flu virus, which might be brought to them by an experimenter. It is particularly important that human cells and animal cells, especially those of birds, are always kept properly apart. Mixed cultures would provide an environment in which a bird flu virus could, by mutation, adapt and be selected for the ability to grow in human cells, and something really dangerous might arise.

In general, even human cancer cells ought not be dangerous in themselves to an experimenter: even if they entered her body through, for example, an accidental cut with broken glass, they should be recognized as foreign by her immune system and killed. Unfortunately, many students new to cell culture decide it would be interesting to culture some cells from their own bodies, for example, from a cheek swab. This would be unwise: if one of those cells mutates and becomes immortal, and gets back into you somehow, it may be indistinguishable from 'self' and you will have no defence against it as it develops, potentially, into a cancer. The risk may be low, but not worth taking.

In vivo issues

Some projects in mammalian synthetic biology involve placing an engineered device, or just biomolecules produced by synthetic biological devices in culture, into a mammal's body. This might be done, for example, to diagnose or to treat a disease. The internal structures and defence systems of mammals can, unfortunately, make this approach very difficult, and it may require very careful planning.

Access

Some parts of mammalian bodies are easy to access: the surfaces of skin, mouth, and eyes are highly accessible and those of the rectum and vagina only slightly less so. Molecules or cells intended to act on them can therefore be applied directly. The main routes of access to other parts of the body are via the gut, as a pill or as adulterated food, or via the bloodstream as an **inoculation**.

Access via the gut is complicated by the fact that the main function of the digestive system is to digest, that is, to break large molecules such as proteins down to their constituents, which will then be absorbed. While small manufactured molecules such as aspirin survive the gut far enough to be taken up and released into blood, proteins do not. This is why even small protein-like hormones such as insulin cannot be given in tablet form and have to be injected. Some viruses can infect the body via the gut, and this may be useful in introducing synthetic biological devices, as long as the virus itself is rendered harmless by the replacement of its normal genetic payload with the engineered device.

Even inoculation may fail to transfer an injected molecule to the site it is intended to reach. Important organs such as the brain and spinal cord are separated from the general fluids of the body by the **blood–brain barrier**, and large 'biologics' such as proteins may well not cross it. Other places, such as the testis, have similar barriers.

Immunity

The other major challenge to the application of synthetic biology products to living animals, especially repeated application, comes from the **immune system**. Mammals have complex, long-lived, resource-laden bodies that would be easy picking for simpler animals if it were not for their ability to defend themselves, and their defensive systems are therefore extensive and complicated.

Some defences of the mammalian body are purely mechanical (tough skin, for example), and some rely on the **innate immune system**. The innate immune system is a set of cell types and proteins that can, collectively, detect the presence of common pathogens and kill them either chemically or by directing the destructive, phagocytic activities of cells such as **neutrophils** and **macrophages** to destroy them. The innate immune system can also detect signals given off by sick or dying cells of the body, and launch an attack on those cells and anything else, such as bacteria, in their vicinity: this attack is the **inflammatory response** characterized by heat, swelling, redness, and pain.

In addition to the innate immune system, mammals have an **adaptive immune system**. This is one of the body's two great learning machines, the other being the brain. When it develops, the adaptive immune system uses very specific gene-mutating mechanisms to generate a huge range of receptors on its cells. Some of these recognize nothing, and the cells carrying them die. Some recognize structures of the body very well, and the cells carrying them also die. This is important—if these cells survived to mount an attack on the body's own tissues, an autoimmune disease would result. This leaves just cells carrying receptors that interact very weakly with the body's own proteins to survive. This interaction is too weak to cause the cells to be activated, but is enough to show that the receptors do work as receptors. By the process just described, the adaptive immune system is therefore equipped with a vast library of cells (i.e. a collection of different versions of the cells) carrying functional receptors for unknown proteins and other molecules.

Cells such as macrophages patrol the body and pick up samples of proteins and other molecules they find there, working especially hard in areas where inflammation is present, and **present** them to the cells in the library. If one of the presented molecules is recognized strongly by one of the receptors, the cell bearing it becomes activated. It multiplies to make more of itself, and (depending on exactly the cell type) it either becomes a lethal cell that will kill any cell carrying that new protein, or it makes antibodies that target killer cells and cell-killing complexes to that new protein, or it helps with these events. When the entities carrying the new protein (bacteria or viruses, for example) have been disposed of, some of the activated cells persist as **memory cells**, ready to launch a very rapid and lethal reaction should the new protein appear again. This is what makes the adaptive immune system 'adaptive'—it learns the molecular signatures of past threats and reacts strongly to them should they reappear. **Vaccination** works by exposing the adaptive immune system to proteins from specific bacteria or viruses, so that it will remember them and launch a fast response should the real disease-causing virus or bacterium appear.

The problem for the synthetic biologist is, of course, that the adaptive immune system is as likely to recognize a new biomolecule as the signature of a possible threat as it is to recognize a virus protein as one, and the second time the biomolecule is introduced to the body it will meet with a rapid and forceful attack.

The problems versus the power

In describing the special features of mammalian systems, this chapter has focused on problems, because these have such a great influence on system design. It would be a mistake, though, to leave the impression that mammalian synthetic biology is all about problems and to be avoided. Many of the problems result from features that can give great power to the synthetic biological approach.

Mammalian systems offer a huge range of ways to control the expression of engineered genes. It is relatively easy to connect to signals in mammalian cells to make engineered systems responsive to natural body signals and to talk back to the body by making other signals. The compartmentalized nature of mammalian cells allows metabolic pathways to be set up in different compartments so that they do not interfere with one another. Systems can also be designed to be activated in just some parts of the body, and even at specific times. Most important of all, learning how to apply synthetic biological techniques to mammals in general is the route to learning to apply them to ourselves.

Chapter summary

- Broadly, mammalian synthetic biology is more complicated than synthetic biology in bacteria: this is a challenge to designers, and at the same time an opportunity as mammalian systems offer multiple, powerful mechanisms of control.
- Mammalian genes are typically split into a number of exons separated by introns, and the introns are spliced out during RNA processing. Alternative splicing gives rise to alternative forms of the protein.
- Gene expression is also controlled by the state of chromatin, in which condensation inhibits expression. Chromatin changes tend to influence entire regions of chromosomes.
- The transcription factors, chromatin-modifying complexes, and signalling systems that collectively control the expression of genes are grouped under the term 'epigenetic'.
- Mammalian cells have multiple separate internal compartments (organelles) in which different cellular functions take place in relative isolation.
- Mammalian cells are fragile and have very specific requirements for culture, which effectively mimic the inside of the body.
- Most mammalian cells isolated from tissue can be grown for only a set number of cell doublings—the Hayflick limit.
- Cell lines and cells treated with immortalizing genes escape this limit, but show properties different from the natural cells.
- Within the intact mammalian body, there are barriers to free movement of introduced molecules, cells, and systems: the blood–brain barrier is an important example.
- The mammalian adaptive immune system learns to recognize unfamiliar molecules, and will mount a rapid and effective attack on them the next time they are introduced. This is a limitation for projects that hope to treat a disease by repeat dosing with a molecule not already made in the body.

Further reading

Allen T, Cowling G (2011). *The Cell: A Very Short Introduction*. Oxford: Oxford University Press.

This book provides a clear overview of both bacterial and eukaryotic mammalian cell structures.

Cold Spring Harbor Laboratory (2011 but maintained indefinitely). DNA from the beginning. http://www.dnaftb.org/.

A free, easy-to-navigate, animated educational resource provided by the Cold Spring Laboratory, which has long been an important world centre for genetic science.

International Union of Immunological Societies (2017 and updated). Immunopaedia. https://www.immunopaedia.org.za/.

An excellent free educational resource maintained for use in the whole world, and based in Africa in the heart of the HIV crisis (AIDS is primarily a disease of the immune system).

Kemp TS (2017). *Mammals: A Very Short Introduction*. Oxford: Oxford University Press.

This book provides an excellent, concise introduction to mammalian life.

Leica Microsystems (2017). Introduction to mammalian cell culture. https://www.leica-microsystems.com/science-lab/introduction-to-mammalian-cell-culture/.

This website, produced as a free educational resource by a manufacturer of microscopes, is a concise, well-illustrated introduction to cell culture.

Discussion questions

2.1 How many ways are there to control the expression of a mammalian gene? Which of these are most useful to synthetic biology?

2.2 Imagine you are involved in a synthetic biology project that aims to make a reliable system to detect whether potential new drugs will damage the cells (endothelial cells) that line human blood vessels, before clinical trials in real humans would put anyone in danger. Assume that a genetic system for reporting cellular damage and stress has already been designed and built, and its builders just need cultured human endothelial cells into which to put it. Also assume that an ethical source of human arteries exists (it does—the umbilical cords that are cut in childbirth). How would you go about setting up a culture of the cells, and what would be the best way to ensure its longevity for many years of use?

2.3 Imagine you have been asked to lead a synthetic biology programme that aims to treat a human disease, caused by a congenitally missing metabolic pathway, by adding enzymes to run a simplified, synthetic version of that pathway inside cells of a human body. Assistants working with you have come up with three ideas. Kim suggests producing the enzymes, in a form that will be taken up by human cells, in culture, and injecting them into the patient every week. Pat suggests engineering a virus, capable of entering human cells, so that the normal genome of the virus is replaced by genes encoding the enzymes: the virus will no longer be able to replicate or to enter the human genome. The virus will have to be given again every year. Sam suggests engineering the virus a different way, so that the genes encoding the enzymes will be placed in a random position in the genomes of infected cells. Which of these approaches would be best, taking into account both safety and efficiency? Can you think of a better alternative?

3

TECHNOLOGIES FOR MAMMALIAN SYNTHETIC BIOLOGY

Dr Leonard J. Nelson and Professor Alistair Elfick

Learning Objectives

- Describe the methods by which DNA can be edited, written, and constructed.
- Give a summary of the challenges faced in authoring DNA.
- Show how the function of our DNA instructions can be lost over time.
- Explain how practitioners are seeking to protect their DNA from corruption.

Synthetic biology aspires to the ability to create, or improve, function in biological systems by considering them as a type of material from which to design. The 'engineering' approach to synthetic biology holds as its key notion the Design–Build–Test cycle (Figure 3.1) and the optimization of each step towards a degree of predictability that allows a single orbit of that cycle to deliver a best design. At the time of writing, we are a significant way from this goal of rational design; indeed, in a material as complex and responsive as mammalian cells it may remain unattainable for some time.

In Chapter 2 we introduced the fundamentals of the 'central dogma' of molecular biology as it pertains to the mammalian cell; that information held as DNA is transcribed into RNA and translated into functional proteins, with this process being regulated at the genetic and epigenetic levels. It follows that the key underpinning

Fig. 3.1 Synthetic biology embraces the engineering principles of Design–Build–Test. Modern engineering design has been refined to the point that products may be the result of a single orbit around the Design–Build–Test cycle. Synthetic biology is far from this point at present. You may see 'Learn' added to this cycle, but we view it as an inherent part of 'Test' so it needs no repetition.

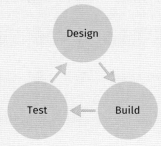

technological tool that enables synthetic biology is the ability to produce genetic devices by modifying or creating DNA code.

An analogy can be made between the history of the published word and that of DNA publishing. In 1493, Johannes Gutenberg, a goldsmith, invented a printing press which has been called one of history's most influential technologies. Prior to Gutenberg's printing press, books were hand-scribed by monks, took years to produce, and were fabulously expensive. Afterwards, literature became accessible, reading became a skill of the masses, and information and thought were more diversely communicated; societies changed, the world changed. Nowadays, desktop publishing is universally available, and while the written word still exists on paper, it has also moved into the Internet.

In Chapter 3 we give an overview of the rapidly developing field of writing DNA because, through our profound new abilities to read, edit, and write DNA, the world is changing.

We can read DNA very well

The investigation of how organisms function at the molecular scale has been an important feature of biology for around a century. In that time, we have established a wealth of knowledge about the genetic code and its operation. DNA, which is common to all organisms, is a universal medium for storing biological information and for a long time some people claimed that, if we could read its nucleotide sequence, we would unlock all the secrets of biology, and of disease. In the latter part of the twentieth century there started a rush towards reading the genetic code base by base, a process called DNA sequencing (see Scientific approach 3.1). Over time, sequencing has become very accurate, very quick, and much, much cheaper. What took years of work in a warehouse full of equipment can now be achieved in a few hours on a USB device you can plug into your laptop. Figure 3.2 shows a historical timeline of advances in DNA sequencing.

Yet an ability to read DNA does not mean that we understand everything that it does. This point has been illustrated well by researchers at the J. Craig Venter Institute. In 2016, they published a paper in which they stripped away all of a bacterium's DNA that they could not connect with a specific biological function. If there were nothing left to learn, then all of this DNA would have been unnecessary, and the bacterium would have lived without it. But the resulting cell could not survive, and the researchers started to put back genes that they had predicted to be non-essential and had to replace about forty per cent of what they had deleted before the bacterium could live. This work gave sobering confirmation of the extent of our shortcomings in our understanding of the functions of the DNA of even the simplest of life forms.

Fig. 3.2 Timeline showing over 100 years leading to the development of DNA (Sanger) sequencing—culminating in the completed first draft of the human genome sequence—and the advent of NGS.

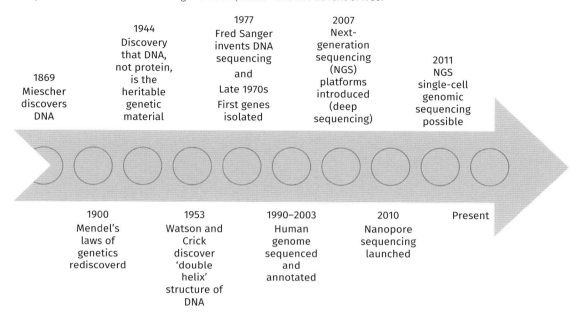

Scientific approach 3.1
Fifty years of DNA sequencing in twenty sentences

First-generation sequencing technologies emerged in the 1970s. The method that rose to early prominence was invented by English biochemist Frederick Sanger (Figure A). The Sanger method synthesized DNA chains by copying a template strand (i.e. the strand to be sequenced), with chain growth halted at each step by random inclusion of one of four possible dideoxy nucleotides. Lacking a 3' hydroxyl group, the dideoxy nucleotide prevented the addition of further nucleotides. A population of truncated DNA molecules was produced, the lengths of which correlated with each of the sites of that particular nucleotide in the template DNA. The molecules were separated according to size in a procedure called electrophoresis, and the nucleotide sequence inferred either manually or using a computer. Later, the method was automated using sequencing machines in which the truncated DNA molecules, labelled with fluorescent tags, were separated by size within thin, glass capillaries and detected by laser excitation. Sanger sequencing has been the most prominent reading technology for the past forty years and is still used today as it has the ability to read quite long strands of DNA, up to 1000 nucleotides.

Figure A shows the Sanger (chain-termination) method for DNA sequencing.

Sanger sequencing has been joined and to some extent superseded by next-generation sequencing (NGS) technologies. NGS is a set of techniques that have a key similarity that distinguishes them from Sanger sequencing; they all sequence many thousands of short pieces of DNA at the same time. Doing this, and also integrating base addition and data readout steps, enables them to be much quicker than Sanger sequencing and much cheaper too. There are different types of NGS from different companies, but they share a first step of cleavage of the DNA into short, contiguous, or neighbouring, strands. These 'contigs' are then anchored to a surface, or bead, for reading. The first step is amplification of a single contig of single-stranded DNA into many identical co-located

Fig. A The Sanger (chain-termination) method for DNA sequencing. (1) **Reaction mixture**—a primer is annealed to a sequence. (2) **Primer elongation**—reagents are added to the primer and template, including DNA polymerase, chain-terminating dNTPs, and a small amount of all four 'natural' dideoxynucleotides (ddNTPs) labelled with fluorophores (coloured). During primer elongation, the random insertion of a ddNTP instead of a dNTP terminates synthesis of the chain because DNA polymerase cannot react with the missing hydroxyl. This produces all possible lengths of chains. (3) **Gel electrophoresis**—the products are separated on a single-lane capillary gel, where the resulting bands are read by an imaging system. (4) **Bioinformatics**—this produces several hundred thousand nucleotides a day, data which require storage and subsequent computational (bioinformatics) analysis.

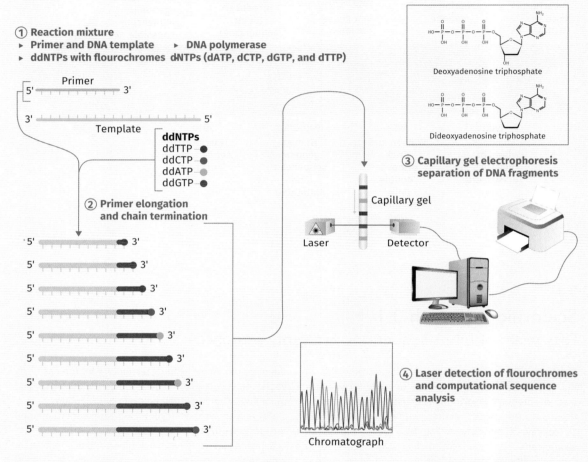

single strands. This ensures that there are sufficient identical molecules such that they can be read. Reading the base order is achieved using sequencing-by-synthesis; when a polymerase adds a nucleotide to the growing strand its type is captured as a read-out (a fluorescence signal, emitted light, or a released proton).

Recently, third-generation sequencing technologies are becoming available. These are based on nanopores created in thin membranes. When a single-stranded DNA strand is pulled through the nanopore it creates a different electrical resistance depending on the type of nucleotide within the nanopore at a specific moment. These approaches are very attractive as they do not need the pre-amplification of the DNA, they can do long reads, and the hardware is very compact indeed.

An excellent historical review of DNA sequencing technologies has recently been published by Heather and Chain (see 'Further reading').

We can edit DNA pretty well

Almost as soon as we began to understand genes and their function, we began to deconstruct them to understand their operation. In the early days, tools were limited and indirect. First attempts at genetic modification predate any understanding of the workings of DNA. Chemicals or radiation were used to randomly mutate genes, leading to extraordinary experiments like the gamma gardens of the early 1950s, where 'atomic gardeners' sought to accelerate evolution of new plant breeds.

This process of random mutagenesis has been used widely in microbes, but it is imprecise and inefficient as researchers need to discard many mutants in the search for the traits they desire (obviously in higher organisms this isn't ethically acceptable). This has motivated the investment of huge effort in moving from the mutation of DNA to its modification.

Rewriting approaches—genome editing tools

Readers will be very familiar with editing documents using a computer; it is easy to add and delete text, 'cut and paste', and even have spelling and grammar checked. Over the past forty years this has become reality, researchers have been developing all this for DNA and, by now, it has become a very precise and reliable technology. Genome editing involves the insertion, deletion, or modification of a selected DNA sequence at a specific site in the genome.

The modern era of genome editing was ushered in by the discovery in the 1960s of enzymes that recognize a specific sequence in DNA and then act as molecular scissors to snip the sugar–phosphate backbones of both DNA strands at, or near, that site. The discovery and characterization of these restriction enzymes (restriction endonucleases), so called as their biological role is to *restrict* the ability of viral DNA to infect a bacterium, won the 1978 Nobel Prize for Werner Arber, Daniel Nathans, and Hamilton Smith.

Technologies for genome editing have advanced greatly in recent years. The identification of additional families of nucleases has provided opportunity for their re-tasking as 'designer nucleases', targeted genomic methods for editing DNA in cells (Figure 3.3a). Genome editing tools include zinc-finger nucleases (ZFNs), and transcription activator-like effector nucleases (TALENs). These are based on engineered restriction enzymes which allow site-specific genome engineering. However, these systems are limited as they usually permit only one genomic edit at a time. In fact, re-engineering of either ZFNs or TALENs is required for each new target. This makes these earlier designer nucleases tricky to engineer, time-consuming, and expensive.

CRISPR/Cas9 revolution

Recent and rapid breakthroughs are now realizing the dream of easy and affordable DNA editing. An extremely powerful technology has been added to the synthetic biology toolbox—the CRISPR system. CRISPR stands for clustered regularly interspaced short palindromic repeat—originally discovered in bacteria as protection against horizontal gene transfer. CRISPR recognizes foreign DNA in bacteria via an adaptive immune system for defence against invading viruses (bacteriophage), plasmid DNA, and transposons (more on these later). CRISPR requires the presence of guide RNA (gRNA) and an RNA-guided endonuclease,

Fig. 3.3 Designer nucleases used in in synthetic biology for targeted genomic editing of DNA in cells. (a) Genome editing tools exploit precision **designer nucleases** including ZFNs, TALENs, and CRISPR/Cas9. In the latter case, a gRNA 'programmes' the Cas9 protein to a particular complementary location on the target (host) DNA. These enzymes cut double-stranded DNA at such user-defined target sites in the genome (genes), creating double-stranded breaks (DSBs) in the DNA. The mammalian cellular machinery can repair the DSB using two major pathways involved in DNA repair called **homologous recombination** (HR; see Figure 3.6 and related text for details) and non-homologous end joining (NHEJ). The NHEJ 'error-prone' pathway can lead to frameshift mutations, and in this example, results in gene disruption, or gene KO (blue square). Technologies for inserting genes also exist (see Figure 3.6, and related text). (b) Schematic showing the CRISPR/Cas9 components for gene KO ('*identify and chop!*'). In the simplest CRISPR/Cas9 genome engineering experiments, the repair of the cut DNA is left to the cell's repair mechanisms without providing an exogenous DNA template. The cell proceeds with NHEJ of the cleaved fragments. NHEJ binds the DSB back together, but in the process may insert or delete nucleotides (indel). This method typically leads to a frameshift mutation and a KO of the targeted genetic element's function. The CRISPR system is discussed in more detail in the text (and in Figure 3.6).

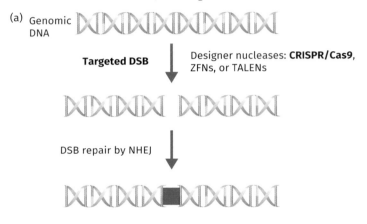

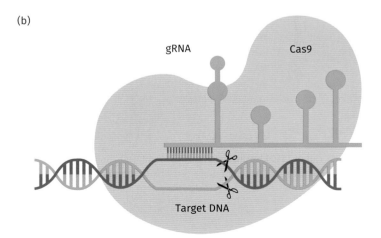

typically one called Cas9. The CRISPR system can insert or delete desired sequences (called indels) at very precise locations along the entire genomic DNA sequence. The gRNA 'programs' the Cas9 protein to a particular complementary location on the target (host) DNA. Cas9 then cuts the desired piece of target DNA, allowing whole genes to be edited. This can include insertion or deletion events, as chosen by the experimenter, for the study of gene function, expression, and inhibition in cells. Perhaps the most common use of CRISPR/Cas9 in genome engineering experiments is knockout (KO) of the targeted genetic element's function as depicted in Figure 3.3 (see also Figure A in Case study 3.2, later). That's not all—CRISPR allows multiplex experiments, in which several gRNAs can be used in one CRISPR array experiment, so enabling simultaneous editing of several genes within the genome. Given that the human genome is around three billion base pairs long, in CRISPR, we now have a genome editing tool which is reliable, quick, cheap, and robust enough to do gene therapy (although at the time of writing there remain important safety concerns: we need to be really sure that the only changes made are those we intended). CRISPR/Cas9 is also highly flexible. For example, it can be used to introduce point mutations into DNA (or RNA) whereby only a *single* nucleotide base is changed, inserted, or deleted. Such gene mutagenesis can be detrimental, as occurs in cancer, or potentially beneficial, as occurs in natural selection.

Transposons are also important genome modifiers in evolution and genome regulation and function and are found in almost all organisms. These 'jumping genes' are segments of DNA that can move around the genome. Depending on where they land, they could generate, or even reverse mutations, although most events are silent—with no effect on the cell phenotype. Geneticist Barbara McClintock of Cold Spring Harbor Laboratory in New York won the Nobel Prize in 1983 for discovering these important genome modifiers in maize.

Assembly of DNA parts

The scale of ambition for DNA editing has expanded in line with our design of increasingly powerful technology. Instead of just modifying a single gene, our attention is moving to making a complete genetic 'construct'. In our text editing analogy, this is moving from cut-and-paste of a word (coding sequence) to a whole sentence (open reading frame), or even a paragraph (a multi-gene construct). Underpinning this, and a foundational concept for synthetic biology, is the assembly of biological parts, often called bioparts, such as Golden Gate, Gibson, Paper-clip, and EMMA, or BioBricks. There are numerous strategies for assembling bioparts: these strategies exploit nucleases. There are yet more assembly strategies that harness alternate DNA editing approaches. All have their strengths and weaknesses, and there are numerous reviews which the curious reader can pursue (see 'Further reading'). We should note that in mammals and plants, non-homologous end joining (NHEJ) is the preferred pathway, while in yeast it is homologous recombination (HR), which is often utilized in DNA assembly and editing approaches.

We can write DNA quite well too

At the same time that people started to sequence DNA, researchers also began to synthesize it. The key breakthrough in this field was the development of a chemical process for writing DNA, a process that remains unsurpassed in its fidelity. In the mid-1960s, Har Gobind Khorana and colleagues pioneered the development of phosphoramidite-based oligonucleotide synthesis, for which Khorana was subsequently jointly awarded the Nobel Prize in 1968 with Marshall Nirenberg and Robert Holley: 'for their interpretation of the genetic code and its function in protein synthesis'. This chemistry allowed single-stranded DNA to be synthesized one base at a time, nucleotide by nucleotide, in the desired sequence (Figure 3.4). As you can see, it is a simple procedure which is easily automated, but it is quite slow, requiring washes between each step. It has excellent fidelity with a reaction efficiency exceeding ninety-nine per cent at each cycle. Good though this is, it does mean that when trying to synthesize a 100-nucleotide long chain, only about thirty-six per cent of the product will be the desired base sequence.

The process has been tuned and optimized for more than forty years. In the early days, only very short pieces of DNA (oligonucleotides) could be written and the costs were high. There has been a concerted effort to reduce cost, mainly through miniaturization and automation of the synthesis process. DNA synthesis is now cheap enough to enable researchers to contemplate experiments that would have been prohibitively expensive a few decades ago (see Case study 3.1).

Fig. 3.4 The oligonucleotide synthesis cycle. The first nucleotide is anchored to a surface and then (1) chemically 'deprotected' (involving removal of the DMT protecting group), which alters it such that the next nucleotide can be added; (2) the act of adding a base stops the addition of any further bases. Any chains which did not have a base added are then 'capped'; (3) which stops any subsequent base addition to an incorrect strand. The process then starts again and cycles around until the complete oligo is produced.

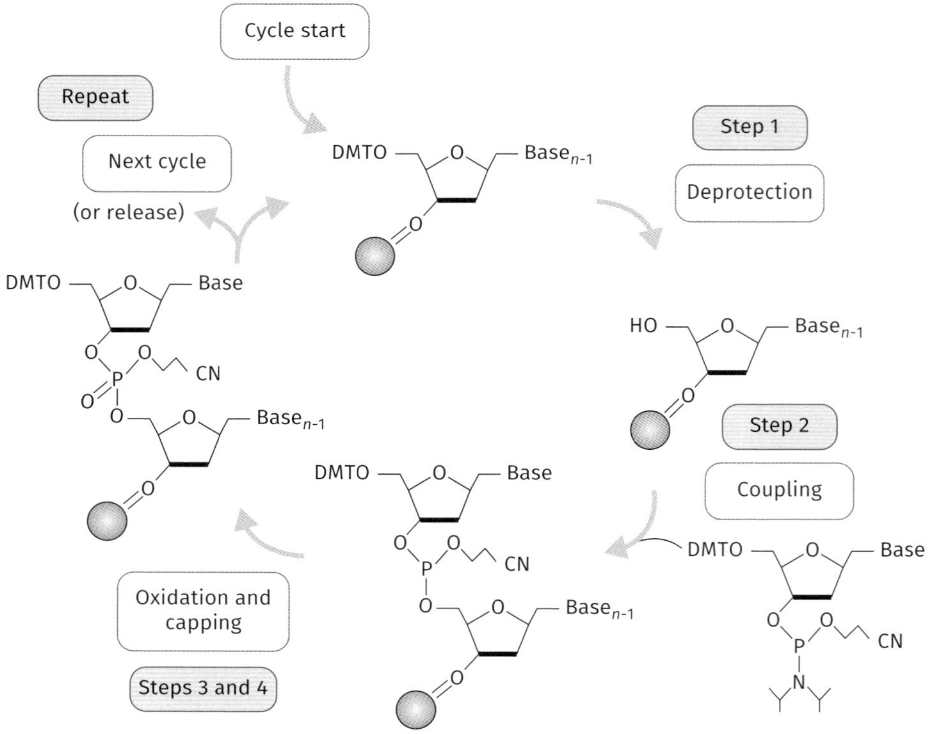

Case study 3.1

While this chapter focuses on tools for mammalian synthetic biology, it is very important to realize that these tools (e.g. CRISPR, Gibson assembly/BioBricks; Sanger sequencing/NGS, etc.) were often initially developed, tested, and optimized in non-mammalian hosts (e.g. bacteria and yeast)—before application of these techniques in mammalian systems. Here, we present a case study utilizing synthetic biology tools in yeast; later (see Case study 3.2) we consider a **gene knockout (KO)** protocol for mammalian cells using a commercially available CRISPR system (see Figure A in Case study 3.2).

Rebuilding the yeast genome by the Sc2.0 consortium

Around the time of publication, an international Sc2.0 consortium will finish the complete resynthesis of the genome of brewer's yeast, *Saccharomyces cerevisiae*. Every single one of its twelve million base pairs, in its sixteen chromosomes, will be replaced by a chemically synthesized alternative, constructed from oligos, but the new genome will also have experienced a set of design changes. The genome will be shrunk though the removal of elements recognized as introns (which are thought, at least in some cases, to have no function), all the TAG 'stop' codons will be swapped to TGA, all non-essential genes will have an inducible recombination site added upstream of their coding sequence, and in the boldest step, all tRNA sequences will be removed from the genomic DNA and added to an entirely new (seventeenth) 'neochromosome'. This technical tour de force will generate a genome possessing unique properties to allow unprecedented questioning of genetics, and the pursuit of industrial yeast strains with enhanced properties.

But our capacity to author DNA remains modest

Composing genetic instructions using bioparts works well for simple statements but becomes challenging very quickly, especially if logical statements are involved. For example, an instruction of '*If* glucose high, *make* adiponectin' (see Chapter 5), is within the bounds of our technical ability. It does not take much, however, to step outside these bounds; linking the previous simple statement to insulin's circadian clock (day–night cycle) would be a big task. And creating a robust oscillator like the circadian clock from scratch is some way in the future.

The explanation lies in our emerging appreciation of the role of systems biology. Even simple genetic statements are executed within a very complicated context, the cell. Our ability to predict how a synthetic circuit will interact with the host chassis is inversely proportional to the circuit size; the more elements in a circuit, the lower our confidence of function as intended. There exist a number of computational tools to assist in the design of a circuit (e.g. CelloCAD: http://2016.igem.org/CelloCad), but these are yet to be fully implemented for eukaryotes. Unfortunately, the toolset for interrogating the interaction of a synthetic circuit with a mammalian chassis is rather limited at present, though the pace of development may quickly change this.

Writing, or synthesizing, DNA is becoming easy but our ability to predict how a novel biomolecule will behave is modest. It's easy enough to recapitulate natural protein domains, append them together, and tweak them a little. However, to design a completely novel, correctly folded, protein that fulfils a predefined function remains very difficult. The challenge lies predominantly in the folding of a functional protein; we can define the desired amino acid sequence but modelling how it will fold holds many challenges. Efforts to address this are ongoing, with initiatives such as the Rosetta software for computational modelling of protein structures having made much progress. Rosetta is enabling advances in *de novo* protein design, enzyme design, ligand docking, and structure prediction for macromolecular complexes.

Another characteristic worthy of note is that chemically synthesized DNA is epigenetically clean; this may be at once both an advantage and a disadvantage. For example, lack of methylation of the DNA will ensure that genes are not repressed through association with a closed chromatin state, but this also interferes with the potential to package large synthetic DNA constructs for delivery by transfection.

The next challenge is getting the modified DNA into mammalian cells

This process is called transfection, in which foreign DNA (or RNA) sequences of interest are introduced into animal cells. Transfection can be used to express new proteins in the host cell (see Chapter 5), or to introduce genes that alter the expression of the natural proteins of the cell.

Transfection methods are numerous, but the most common technologies used are outlined in Figure 3.5. Non-viral methods include electroporation and lipofection: Electroporation is a very widely used physical method of delivering nucleic acids into cells by direct transfer. Electrical pulses delivered to cells are thought to create transient pores in the cell membrane, allowing the DNA to pass into the cell. Lipofection is a highly efficient chemical approach and is at the time of writing the most common gene transfer method (Figure 3.5b). The DNA cargo, which is often carried on plasmids, is easily transferred across the cell membrane in special spherical microvesicle carriers, called liposomes, by endocytosis. More exotic carrier methods for introducing DNA into cells are emerging, including the use of gold nanoparticles, self-assembling protein cages, and optical transfection using optical tweezers. Crucially, as with *any* transfection procedure, there are limitations and sometimes quite extensive optimization is required. As such, careful consideration must be given to the various steps required to achieve the best transfection efficiency. These include optimization of the cell type (e.g. continuous cell lines vs primary cells), choice of transfection technology (size limit of packaged DNA, low vehicle toxicity), use of high-viability cells, growth conditions (cell confluency, trophic factors, medium composition), and type and purity of delivery vehicle (plasmid DNA being the most common). Another feature to consider is the stability of the transfected nucleic acid payload following entry into the cell. This is discussed in the next section.

Getting the DNA to stay in the cell can be difficult too

Following successful transfection, there will be a mixed population of cells exhibiting stable or transient transfection. The latter is much more likely to be observed, and in this the foreign DNA remains independent of the host genome,

Fig. 3.5 (a) Gene transfection technologies used in synthetic biology. Gene transfer into mammalian cells can be achieved by diverse means. Three of the most common general methods are shown (with examples): non-viral methods (electroporation and lipofection), and viral integration (lentivirus). (b) Typical mammalian cell transfection technique: lipofection. Liposomes are spherical vesicles that can carry cargo (DNA or drugs). Liposome-mediated transfection of plasmid DNA is shown using liposomes as delivery vehicles (liposome section is shown). Liposomes (lipofection reagent) and mammalian cell membranes both have similar phospholipid bilayer structures, enabling the DNA–lipid complex to easily cross into the cell by endocytosis. Once inside, DNA is released into the cell cytoplasm and eventually can integrate into cellular genomic DNA in the nucleus; permitting transient or stable gene modification.

(a)

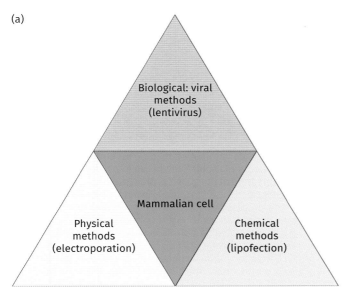

(b)

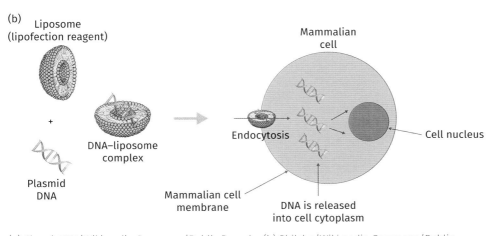

(a) Chemist234/Wikimedia Commons/Public Domain. (b) Philcha/Wikimedia Commons/Public Domain.

and the DNA and the transgene products produced from it are diluted by cell division, and effectively lost. Stable transfection is the *direct* integration of foreign DNA into the host cell genome. Only about 1/10,000 cells undergo stable transfection, so special selection procedures are needed to select the stably transfected transgenic cell population. Antibiotic selection is used as a common method for the identification of stable selection in the cells. The antibiotic resistance gene (e.g. puromycin resistance) is inserted into the expression vector. Only cell clones carrying both the antibiotic resistance gene and integrated transgene will survive. There is also a risk that transgenes will be subject to epigenetic silencing. To defend against this, plasmids can be designed to have a high probability of integrating into a safe harbour site in the target cell genome (Figure 3.6). This technology is used often for gene knock-in, typically for gene correction strategies. Safe harbours protect the integrated gene of interest from transgene silencing and permit greater control over transgene expression.

Transient transfection is still a useful approach for initial testing to see if the desired transgene has been incorporated into the plasmid, and is working.

Fig. 3.6 Safe harbour strategy to enhance stable transfection in mammalian cells: gene insertion or gene knock-in technology. Safe harbour is a method for ensuring stable integration of the transgene (from a donor plasmid) into the cell into a desired locus, and subsequent translation and expression of the gene's protein product. Site-specific genome editing tools such as CRISPR can be used to initially insert the desired gene by **homologous recombination**. This process creates DSBs in the genomic DNA. The donor plasmid DNA containing the desired 'knock-in' insertion sequence (DNA) is also transfected into the cell—which is incorporated into the safe harbour site through homologous recombination. The knocked-in gene can now be analysed for expression for its corresponding protein product. Note the plasmid contains three essential elements— promoter region (green and grey box with arrow) which drives expression of the inserted gene (orange), while the presence of an antibiotic resistance gene (blue) allows selection of the desired cell clones.

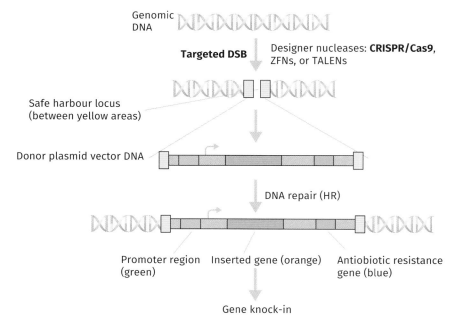

Genomic DNA

Targeted DSB   Designer nucleases: **CRISPR/Cas9**, ZFNs, or TALENs

Safe harbour locus (between yellow areas)

Donor plasmid vector DNA

DNA repair (HR)

Promoter region (green)   Inserted gene (orange)   Antiobiotic resistance gene (blue)

Gene knock-in

Production of stable cell lines provides very useful models for studies of gene function, disease-causing transgenes, and drug compound screening.

Mammalian cellular repair machinery

Essential to understanding gene-editing technologies is the harnessing of the cell's DNA repair machinery, which corrects (with extremely high fidelity) harmful breaks that occur on *both* strands of DNA. Mammalian cells use a highly conserved pathway called homologous recombination (HR) to accurately repair double-stranded breaks (DSBs). This is a crucial process as DSBs—whether endogenous, due to metabolism-induced DNA damage, or exogenous, due to ionizing radiation or chemotherapeutic (anticancer) drugs—are potentially lethal lesions. HR uses an undamaged DNA template to repair the break, leading to the reconstitution of the original sequence.

In a nutshell—CRISPR KO experiments utilize NHEJ (Figure 3.3), while knock-in and safe harbour strategies utilize HR (Figure 3.6).

Finally, we need to know our modified DNA is functional

Having achieved successful transfection of the desired sequence of DNA (or RNA) and enhanced our chances of creating a stable cell line, by using antibiotic selection, we must verify that we have generated the desired sequence at the targeted genomic site. This is most often achieved by **genomic sequencing**. NGS is used for large-scale projects sequencing hundreds of genes or the whole genome. Most often Sanger sequencing (as described in Scientific approach 3.1) is used for transfection verification, as it generally deals with sequencing of one gene at a time.

To understand better why we use genomic sequencing, let's revisit our CRISPR/Cas9 genomic editing technology (see Figure 3.3 and Figure 3.6). Imagine we wish to do a gene KO experiment. Our goal is to KO a particular gene called *ABC* in a human cell line, using a gRNA. Using the CRISPR system described previously, the gRNA acts as a cellular 'GPS'—guiding the Cas9 to chop the DNA at the desired genomic site. This snipping of the double-stranded DNA creates indels in the cell population because the damage to DNA is repaired by an error-prone pathway for DNA damage repair that is always present in mammalian cells. We can harness such indels generated at the targeted genomic site (gene *ABC*), to achieve our desired KO. This is because such edits (i.e. indels) can occur *within* a protein coding region or exon. These can lead to a frameshift mutation (change in the codon reading frame) that will mostly likely terminate protein function—so creating a cell line containing a knocked-out *ABC* gene! Indels can also lead to generation of **nonsense mutations** or **premature stop codons** to produce such functional gene KOs.

Therefore, we must next use **Sanger sequencing** (Figure A in Scientific approach 3.1) to identify all indels generated at the targeted genomic site (gene *ABC*). This enables us to evaluate the editing (targeting) efficiency in our CRISPR KO experiment. It is also important to confirm that the phenotype of the transfected cell is as expected. Phenotyping can include functional properties of the cell following genomic editing, including morphological and physiological, as well as biochemical properties of the cell. Microscopy is used as a core technique to assess cell morphology, including various types of light microscopy, both at modest magnification to observe gross cellular morphology and high magnification to view specific proteins in cells labelled with a fluorophore (a chemical label which fluoresces after being excited by light of an appropriate wavelength). Biochemical properties of the cell can be assessed using a vast array of techniques, using widely available assay kits (e.g. testing cellular ATP content for cell viability). More sophisticated approaches to assess the impact of a gene KO on cells include the use of omics technologies. Omics includes genomics (including genomic editing/sequencing), but also proteomics, lipidomics, and metabolomics. The latter rely on **mass spectrometry,** an analytical technique which can detect changes in small-molecule metabolites (metabolomics), or changes in the entire protein or lipid content of the cell (proteomics). Omic approaches generate huge amounts of data (i.e. terabytes of DNA, or RNA sequences; and metabolite flux data) and require complex data interrogation using bioinformatics software, which is constantly being refined.

Connecting events occurring at the level of DNA and RNA with changes in cellular metabolism and protein expression is seldom straightforward. This is one reason that we find it so difficult to design novel proteins, for specific tasks, from scratch. There has been over forty years of investment in analytical technologies around DNA, thousands of papers, and numerous Nobel Prizes. Yet the technologies for phenotypic characterization of cells have not advanced at the same pace. The next few decades will see this change and, with a bit of luck, help to fill in those gaps in our knowledge of biology's 'dark matter'.

Many technologies for mammalian synthetic biology are now available or under development, to try and fill these gaps. Case study 3.2 is a 'real-world' example of how many of the tools described in this chapter are integrated to achieve a common goal in synthetic biology—gene KO in mammalian cells.

Case study 3.2

Gene knockout protocol for mammalian cells using a commercially available CRISPR system

As we have seen, gene editing with the earlier designer nucleases (ZFNs, TALENs) are difficult to engineer and optimize, expensive, and time-consuming; often taking many months to complete and with no guarantee of success. Streamlined, commercially available systems, such as *CRISPRevolution*™ are now capable of delivering highly-efficient gene KO in only six weeks.

CRISPR editing experiments for mammalian cells begin with culturing the desired cell type (e.g. human HEK293 cell line) and empirically determining the number of cells required (typically, 100,00 cells/well in twenty-four-well culture dishes), the amount of Cas9 protein, and the ratio of gRNA:Cas9—and a suitable transfection method for

the chosen cell type. The main steps in a CRISPR editing gene KO workflow (Figure A) are as follows: (1) **preparation for editing**: design, order, and test primers for genomic polymerase chain reaction (PCR) (i.e. genomic DNA at the region targeted by the single gRNAs will need to be PCR amplified for indel identification (see step 4). (2) **Ribonucleoprotein (RNP) formation and transfection**: the essential CRISPR machinery (gRNA and Cas9 protein) can be delivered to the target cell in one of the three formats: RNP complex, plasmid DNA, or *in vitro* transcribed RNA. Here, we use RNP. (3) **Analysis of editing efficiency**: extraction of genomic DNA is performed in preparation for amplifying the genomic region of interest by PCR. Following amplification, gel electrophoresis of the PCR product is performed to verify amplification of a single band of the correct size. PCR purification of the products is required prior to Sanger sequencing. Following Sanger sequencing, CRISPR edit data require the help of robust bioinformatics software tools, such as **ICE** (inference of CRISPR edits: https://ice.synthego.com) to determine, for example, editing efficiency. (5) **Clonal isolation of KO**: finally, stringent techniques are employed to isolate and verify that we have a newly edited clonal cell line with the desired gene KO—culminating in activity assays and immunoblots (e.g. Western blot) to test the effects of the targeted modification on cell protein expression and functionality. Note that, in this example, CRISPR's ability to knock out gene function relies on the cell's NHEJ mechanism for repairing DNA DSBs (see Figure 3.3 and Figure A, and text for details). The reader is referred to the Synthego website for further information (https://www.synthego.com).

Fig. A Schematic for a gene KO protocol and timeline using the commercially available Synthego CRISPR system. This protocol is used in a variety of cell types, including mammalian cells, as a genome engineering Design → Edit → Analyse workflow. The human embryonic kidney cell line (HEK293) is widely used in synthetic biology experiments as it is easy and cheap to culture in the lab. The four main stages of producing a stable gene KO clonal cell line are shown in bold (see text for details); and the experimental timeline in green. Single-guide RNA (sgRNA) and Cas9 are introduced into cells as ribonucleoprotein (RNP) complexes. Following Sanger sequencing (see text), CRISPR edit data requires the help of robust software tools, such as ICE (https://ice.synthego.com), to determine, for example, editing efficiency.

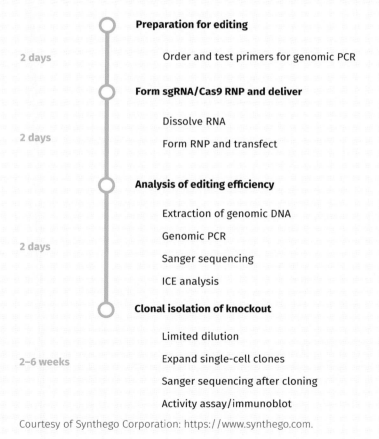

Preparation for editing

2 days

Order and test primers for genomic PCR

Form sgRNA/Cas9 RNP and deliver

Dissolve RNA

2 days

Form RNP and transfect

Analysis of editing efficiency

Extraction of genomic DNA

Genomic PCR

2 days

Sanger sequencing

ICE analysis

Clonal isolation of knockout

Limited dilution

Expand single-cell clones

2–6 weeks

Sanger sequencing after cloning

Activity assay/immunoblot

Courtesy of Synthego Corporation: https://www.synthego.com.

Chapter summary

- Synthetic biology couples engineering principles of Design–Build–Test with molecular biology techniques, to create, or improve, function in biological systems by considering them as a new type of material from which to design.
- Synthetic biology relies both on the 'central dogma' of molecular biology—DNA makes RNA makes protein—and requires some knowledge and understanding of techniques borrowed from biochemistry and genetics.
- A wide range of technologies for mammalian synthetic biology have been developed around these core disciplines to facilitate the redesign of natural biological systems, or ultimately the rational design and fabrication of new biological entities.
- DNA (and RNA) sequencing are fundamental technologies which underpin synthetic biology. Sanger sequencing and NGS are used to capture the precise sequence of nucleotides, such as ACGT (DNA) or ACGU (RNA). Short sequences of DNA (or RNA), single genes, or entire genomes can be sequenced with great accuracy.
- Genome editing tools have progressed from use of simple single-cut restriction enzymes to precision designer nucleases, including ZFNs, TALENs, and CRISPR/Cas9.
- Applications of synthetic biology include gene therapy in humans, enhancement of crop production and disease resistance, and in bioproduction of proteins for industrial and therapeutics, including antibodies.
- As an editing tool, CRISPR/Cas9 may be a game changer for synthetic biology in achieving these goals. It is currently the cheapest, quickest, and most flexible gene editing technology.
- Transposons (jumping genes), present in almost all organisms, and are important genome modifiers in evolution and in genome regulation and function.
- Transfection is the key technology for introducing edited nucleic acids into cells. Chemical (lipofection) or physical (electroporation) methods are most often used to allow entry of plasmids (carrying the desired gene sequence) across the plasma membrane.
- Production of biologically engineered stable cell lines requires stable transfection through direct integration of foreign DNA into the host cell genome. Strategies to select for and enhance for such stable cell lines include antibiotic selection and use of safe harbours.
- Phenotypic profiling of candidate stable cell lines is performed to test the efficiency and functionality of the cloned cell line.

 Further reading

Ellis T, Adie T, Baldwin GS (2011). DNA assembly for synthetic biology: from parts to pathways and beyond. *Integr Biol (Camb)* 3, 109.

A good review of building genetic circuits from DNA parts.

Heather JM, Chain B (2016). The sequence of sequencers: the history of sequencing DNA. *Genomics* 107, 1–8.

If you want to know more about the history of DNA sequencing.

Meneely P, Dawes Hoang R, Okeke IN, Heston K (2018). *Genetics: Genes, Genomes, and Evolution*. Oxford: Oxford University Press.
Excellent general textbook for background and further reading into many of the topics covered in this chapter, including the central dogma of molecular biology, CRISPR technologies, and genome organization, structure, and variation. Interesting also from a genetics perspective.

 Discussion questions

3.1 (This question is designed to prompt some quick web-searching for further information.) A student wanting to make a small change to a gene in a human cell line is unsure about whether to use TALENS or CRISPR/ Cas9 and has come to you for advice. What would you say would be the advantages and disadvantages of each method? Which would you choose? Would your answer be different if the cells were to be introduced into a living human rather than just used in culture?

3.2 You are reviewing the manuscript of a research report that describes the replacement of a particular human gene in cultured cells with a different gene (the promoter, etc. being the original—only the gene has changed). What is the minimum set of controls you would like to see before you are convinced that the experiment has been done properly?

3.3 The Sc2.0 consortium is reconstructing the yeast genome, replacing all TAG 'stop' codons with TGA 'stop' codons. To what uses might an organism with no TAG stop codons be put (in terms of future synthetic biology projects intended to modify the organism further), for which the normal organism could not be used?

4 MAMMALIAN SYNTHETIC BIOLOGY AS A RESEARCH TOOL

Professor Jamie A. Davies

Learning Objectives

- Outline the virtuous cycle in which basic science informs the design of synthetic systems, and new synthetic systems can be used to discover more basic science.

- Describe the two broad ways in which synthetic biology is applied to basic science: as a discovery tool and as a means of testing theories.

- Explain how synthetic biological techniques can allow experimenters to use light to activate specific neurons in an animal brain.

- Define RASSLs, and describe how they can be used to dissect complex signal pathways.

- Outline the strengths and weaknesses of the following methods of hypothesis testing: gene knockout, computer modelling, and modelling by synthetic biology.

- Define the concept of a functional motif.

- Give an illustrated example of a synthetic biological realization of a gene network motif.

- Outline an example of the use of synthetic biology to explore roads not taken in natural evolution.

Synthetic biology is often presented in the context of biotechnology or medicine, where it is a means to achieve a technological end. It may, for example, be a means to produce biofuels or drugs, or be the basis of a device to diagnose illness, or be a method to produce novel tissues to repair a damaged body. All of these applications have the potential to have far-reaching future impacts on society and human well-being, but synthetic biology can also be used to feed back on the process of scientific discovery. This creates a **virtuous cycle**, in which new knowledge can be used to design new synthetic biological systems, and some of these systems can be used as **tools for discovering new knowledge**.

In this chapter, we will present some examples of the ways in which synthetic biological devices can help scientists probe more deeply into the fundamental mechanisms of life, and verify the theories they construct. There are many more cases than we can cover here, and our aim is mainly to give an idea, from a limited number of carefully chosen examples, of the range of ways in which synthetic techniques can be used.

Synthetic tools for analytic biology

Great leaps forward in biological understanding have often followed the invention of new instruments and ways of working. The invention of the optical microscope, of sensitive electrical measuring instruments, of the electron microscope, and of genetic engineering, were all followed by the application of these techniques to research and an almost immediate explosion of new knowledge. In each case, the new technology allowed existing questions to be answered and it also raised new questions, provoked by new features of life that were being seen for the first time.

Synthetic biology is providing analytic biologists with another new tool. Synthetic biological devices offer biologists an unprecedented ability to make very precise changes in cells, effectively to talk to them in their own language and to read their responses. This idea is being adopted so quickly that it is passing beyond specialist synthetic biology journals and is appearing instead in mainstream publications of physiology, neuroscience, genetics, and so on. In the rest of this section, we present a few selected examples to illustrate how synthetic biological tools are being used to deepen our knowledge of how living things work.

 Key point

New technologies arising from scientific discoveries are often first applied to science itself, leading to further discoveries and, eventually, yet more new technologies. Synthetic biology is already being used this way.

Exploring mechanisms of addiction

Addictive behaviour, whether to unnatural stimuli (e.g. drugs, solvents, and tobacco), or to activities that are in themselves natural (e.g. eating and sex) but are done to excess or under inappropriate circumstances, is a major problem to societies the world over. It damages health, it damages relationships, it damages economic productivity, and, in countries where the substance or behaviour in question is not available cheaply and legally, it drives addicts into criminal activity. It is striking that, despite the huge variety of things to which people may become addicted, **addiction itself follows a similar course.** This has led neuroscientists to speculate that there is a common mechanism within the brain, into which these diverse addictive stimuli feed.

Years of painstaking analysis have suggested the critical importance to addiction of neuron-to-neuron signalling, in a particular part of the brain, and using the neurotransmitter dopamine. Dopamine is often released as a result of a mammal experiencing something rewarding, such as tasty food. According to this model, if cells in a region of the brain called the **nucleus accumbens** receive dopamine released by other neurons in other parts of the brain, they increase their expression of a protein called **ΔFosB**. ΔFosB is a component of a transcription factor that activates other genes, some of which cause long-term changes in brain physiology. These changes include an **increased drive** to perform the behaviour (taking a drug, eating energy-rich food, etc.) that resulted in the dopamine release (Figure 4.1). In normal life, this is a natural part of the brain's learning what to do to gain pleasurable outcomes. It will, for example,

Fig. 4.1 A hypothetical model for addiction, in which signals from parts of the brain affected directly by stimulants send dopamine to the nucleus accumbens. Here, dopamine activates ΔFosB, which causes long-term changes in signal pathways in the brain, and alters behavioural drives, generally strengthening drives that resulted in the stimulus.

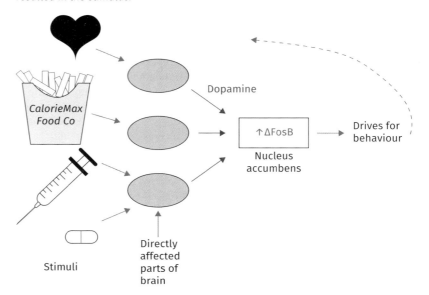

stimulate an animal to engage again in activities such as grooming, court-ship, or lifting logs that might conceal edible invertebrates. When over-driven, though, it will cause changes in the brain that cause an **inappropriately strong or frequent drive** to the behaviour and that will, ironically, make each episode less rewarding than the original, driving an individual to try ever harder to gain the neural reward that was gained the first time.

If this model is correct, then addiction is not really to the drug or behaviour, but to the strong dopamine signalling it elicits. In that case, driving dopamine release directly by some purely technical means, such as pressing a button, should be as addictive as any drug. In 2015, Vincent Pascoli and colleagues tested this idea by constructing a synthetic biological device that would cause neurons to 'fire' and release dopamine molecules across synapses to other neurons in response to light. The link between light and neural firing was achieved using a modified **channelrhodopsin**. These transmembrane proteins are found naturally in algae, and they make pore-like structures in the mem-brane. In the dark, the pores are closed but a light-sensitive part of the mole-cule, derived from vitamin A, will alter its shape and cause the pore to open in the presence of light. When the pore is open, small, positively charged ions (e.g. Ca^{2+}) can pass into cells, altering the voltage across the plasma membrane. The algae from which channelrhodopsins come use this effect to orientate their swimming towards light.

The experimenters introduced their construct, containing a channelrhodop-sin modified and optimized for their purpose, into a specific area of the mouse brain. This was the **ventral tegmental area**, the neurons of which project axons to the nucleus accumbens. Optical activation of the ventral tegmental neurons would therefore result in strong dopamine signalling to the nucleus accumbens neurons. The experimenters also implanted an optical fibre into the brains of the mice, to bring light from a small laser into the ventral tegmental area. They then placed the mice in a living area that contained a lever that, when pressed, would cause the laser to deliver a burst of light pulses. Other levers, not connected to the laser, were also present (Figure 4.2).

The mice were limited to eighty doses of light per day, after which even the previously effective lever did nothing until it became active again the next day. The limitation was a wise precaution, because the mice quickly became so obsessed with pressing the light-activating lever that they reached their eighty-press limit within an hour. They ignored the other levers. Once the animals had become obsessed with the lever, the experimenters introduced a floor that gave the animals a painful, but not dangerous, shock when they pressed the lever. Some quickly learned to avoid the lever but others pressed again and again, even though they would be shocked each time, so important was it to them to get their reward. This was, of course, a crude analogue of humans pursuing compulsive behaviour even at the cost of losing their health, their relationships, their jobs, their liberty, or even their lives. Continuation of behaviour even in the face of negative consequences is the hallmark of addiction.

This experiment added significant evidence that the dopamine-in-nucleus-accumbens theory of addiction is correct. Unlike experiments using dopamine-mimicking or dopamine-blocking experimental drugs, it did not affect the whole brain but **stimulated only very specific neurons**. This specificity allowed it to be much 'cleaner' (free from side effects) than drug experiments could be. It illustrates the way in which synthetic biological devices allow researchers to conduct much more precise experiments than they could before. It may also, one day, help in developing better treatments for addictive behaviours. As a footnote, ethical questions about using living animals for this kind of experiment is part of the general ethical discussion that will be the topic of Chapter 7, which is why these questions have not been discussed here.

Fig. 4.2 The design of Pascoli's experiment that used light and a synthetic biological module to drive dopamine release to the nucleus accumbens directly, when the mouse pressed a specific button.

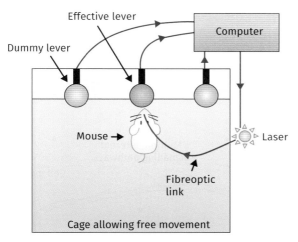

Dissecting cellular signalling pathways

A common feature of cell-to-cell signalling is that, when an extracellular signal is received by a receptor on a cell's plasma membrane, that receptor can activate multiple signalling pathways within the cell, each of which produces a different aspect of the complete response (Figure 4.3). Understanding which parts of the response arise from which pathway is important to basic science, because it is all part of the explanation of how the body works. It is also very important to the development of new medicines because, generally, developing a drug that blocks only the precise internal pathway needed for a cell to do something has far fewer side effects ('off-target effects') than developing a drug that blocks many pathways at the same time. A clear example of this is provided by an important medical challenge: how to stop cancer cells metastasizing (spreading) from the tumour in which they originally develop to other sites in the body. It is metastasis that makes some cancers so dangerous and also makes them hard to reach surgically.

Clinical scientists have known for some time that **breast cancer cells** spread by going first from breast ducts to local lymph nodes, and then flowing in lymph and blood to the rest of the body. Careful comparisons of spreading breast cancers with breast cancers that do not show any spread indicated that spreading cells carry in their membranes a signal receptor called **CXCR4**. Normal breast cells do not make CXCR4: its production by cancer cells is presumably a consequence of genetic mutation in those cells. The normal biological function of CXCR4 is to help cells of the immune system to navigate the body: the receptor detects signals (in the form of the SDF-1 protein) coming from lymph nodes and allows immune cells to travel to them, which they need to do in the course of their usual patrol duties. Clearly, having a breast cancer cell express a receptor that immune cells normally use to find lymph nodes suggests that CXCR4 is what allows spreading cancer cells to find lymph nodes too. This idea was first confirmed by crude experiments in mice, in which all **CXCR4**

Fig. 4.3 It is typical, in mammalian cells, for receptors for extracellular ligands (signalling molecules) to trigger multiple signalling pathways inside the cell, each controlling a different aspect of the response.

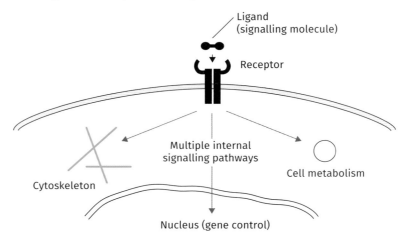

activities were blocked. This greatly **reduced the spread of breast cancer cells** to lymph nodes but, alas, this treatment had very serious side effects on bone marrow cells, which also use CXCR4. Given that CXCR4 can activate many intracellular pathways (Figure 4.4), researchers hoped that one of these pathways might be essential for CXCR4 to help cells to home in on lymph nodes, but not essential for CXCR4 to support normal bone marrow cell function. If that is the case, then there may be a chance of developing a drug that blocks breast cancer metastasis without causing serious side effects. It was therefore important to discover which internal signalling pathways link CXCR4 to lymph homing, and that is where synthetic biological devices proved to be very useful.

There is a class of synthetic biological proteins known collectively as receptors activated solely by synthetic ligands, or **RASSLs** for short. They are usually produced by combining part of a gene that encodes a natural receptor with part of a gene that encodes a receptor for a ligand not present in normal mammals, in such a way that the ligand-binding part of the natural receptor is replaced by one that binds the synthetic ligand, but not the natural one. The internal parts of the natural receptor are still present, and are still able to signal to the rest of the cell. To understand how RASSLs were used for the CXCR4 problem, it is necessary to examine the partners of CXCR4 in more detail.

CXCR4 is an example of a **G protein–coupled receptor (GPCR)**. These membrane-spanning receptors exist in a complex with internal G proteins, which have α, β, and γ subunits (Figure 4.5). When a GPCR binds its ligand, it causes guanosine triphosphate (GTP) to be transferred to the G proteins, and this allows the α subunit to leave the complex and interact with other cellular proteins, triggering internal signalling pathways. There are several versions of GPCR subunits (e.g. $G_{\alpha s}$, $G_{\alpha i/o}$, $G_{\alpha q/11}$, and $G_{\alpha 12/13}$), each with its own preferences for which pathways to activate.

Hiroshi Yagi and colleagues introduced synthetic, RASSL versions of G-protein subunits into breast cancer cells. These could not be activated by activated CXCR4, but could be activated by the artificial compound **clozapine-*N*-oxide (CNO)**. They then tried various combinations of these RASSL versions of G-protein subunits in their cells. One combination, in which the experimenters

Fig. 4.4 Stimulation of CXCR4 triggers multiple intracellular signalling pathways, potentially by interacting with different G proteins.

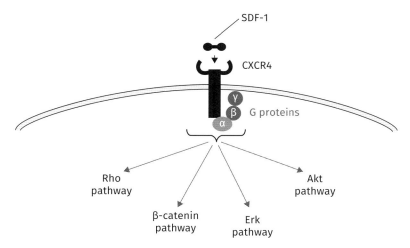

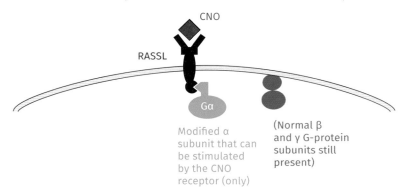

Fig. 4.5 The RASSL form of the G_α protein, which could be triggered by CNO but not by CXCR4. The modified G_α protein, shown in green, that worked was $G_{\alpha12/13}$.

introduced $G_{\alpha i}$-RASSL *and* an engineered form of $G_{\alpha12/13}$ that could be activated by $G_{\alpha i}$-RASSL, was very interesting: the **cells migrated towards sources of CNO** in a dish, as cancer cells normally migrate towards lymph nodes. $G_{\alpha i}$-RASSL was not itself enough—the modified $G_{\alpha12/13}$ was necessary, too, indicating that this precise G-protein subunit is critical for spread of the breast cancer cells with which they had been working. Thus one small step was made in understanding cancer, and in understanding exactly what pathway component needs to be targeted by an anti-cancer drug of the future. The need for painstaking work like this is the reason that treatments for cancers take so long to develop but, without the work being done, there would be no improved treatments at all.

 Key point

In both this example, and the addiction study described earlier, synthetic biology allowed experimenters to connect an artificial stimulant to one very precise point in a complex pathway in a way that cannot be done as easily with drugs.

Building to test ideas

Biologists engaged in traditional, analytic science face a very large problem: **living things are very complicated.** Much of this complication arises from the mechanisms of **evolution.** Genomes are selected according to how well the systems that they encode work to keep an organism alive and breeding: there is no selection for simplicity and elegance, as there is in human-led design and engineering. Thus evolved living things are full of adaptations, of adaptations, of adaptations, and, at least at a molecular level, they appear to be a complicated mess.

Biologists deal with this problem by analysing the messy realities of many organisms and trying to **extract abstract descriptions and principles** from what they see, the real biological systems being regarded as specific instances of the principle. For example, physiologists studying the detailed mechanisms by which the value of a biological parameter (body temperature, body pH, concentration

Fig. 4.6 The architecture of a negative feedback loop. The achieved value of some parameter (e.g. body temperature or blood glucose) is sampled and compared to the ideal 'set point': the difference between the current value and the set point is used to regulate a process (e.g. shivering or insulin production) that will adjust the value to approach the set point.

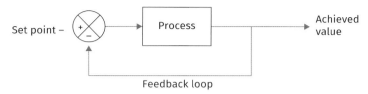

Feedback loop

of blood glucose, etc.) is maintained in the face of potential change created the abstract concept of a **negative feedback loop**, which seems to capture the essence of each of the complicated systems (Figure 4.6). Textbooks on physiology, embryology, ecology, and many other -ologies are full of abstract principles like this but we are left with a big problem: how can we test that the abstract principles we have extracted from messy reality are correct?

There have so far been three approaches to testing the validity of abstract models. The first is to try to intervene in an experimental system to alter the way the system should work, and see if the intervention has the expected result. This is a good way of doing science but, in biology, intervention usually means breaking something, for example, by mutating or removing a gene, or by inactivating a protein. Such experiments therefore prove particular genes and proteins are **necessary** to a system, but are not very good at verifying that the system works as we imagine it does. The second approach is to make a **computer model** of our abstract idea, and ask whether the idea is capable of producing the function we see in real life. This is a powerful way of working and has already added much to our understanding of biology (easily enough for a book on this topic alone—there is no space here to cover properly how useful computers can be in biological research). Computer modelling does, however, suffer from one major flaw that limits its use: computers are 'perfect' places, free from the random thermal noise and the mutual interference of different cellular systems as they compete for space, for materials, and for energy. A programmer can choose to simulate these things, of course, but at the moment we understand them so little that the simulation would largely be guesswork based on yet more abstract models. Therefore, showing that a principle works in a computer-simulated cell does not guarantee that it can work in the 'controlled chaos' of real life (much as economic ideas that perform wonderfully in computer simulations of society have a poor track record of success when tried on real people).

The third, and newest, approach is to use synthetic biology to test the validity of principles drawn from natural biology. To do this, synthetic biologists take the principle and use it to design the **simplest synthetic biological realization** of it that they can, and then see if it works in the environment of a real, living cell or organism. To be clear, this will still not prove that the principle correctly describes an aspect of natural biology, but it does prove that the principle *can* work in real life. It even allows experimenters to explore **evolutionary 'might-have-beens'**, ways of performing a function that could work, but for which evolution actually produced a different solution. In the rest of this chapter, we will use three real examples to illustrate how synthetic biology is being used to test the validity of ideas drawn from real life. One of them also illustrates the idea of exploring evolutionary 'might-have-beens'.

Motifs in gene networks

Analysis of the natural genetic systems used by organisms to process and integrate information about their environments and internal states have suggested the existence of functional motifs—particular ways of organizing genes that act as 'modules' within larger systems. When the genetic systems of diverse organisms are examined, the same motifs turn up again and again, their presence correlating with particular assumed functions in the network. It is not that the same genes are used in every system; rather, it is the same **arrangement** of activating and inhibitory links. These observations are the basis of theories about how gene networks process information. The realism of the theories have been tested in computer models but, as noted earlier, the fact that a motif works in the carefully controlled environment of a computer simulation is no guarantee that it will work in the tumultuous environment of a living cell. This can be tested best by building a synthetic biological version of the network.

An example of this type of experiment is provided by the work of Beat Kramer and Martin Fussenegger, who used synthetic biology to construct a genetic module of a type thought to confer hysteresis in the response to a signal. Hysteresis is a phenomenon in which the threshold of a response to a signal is different according to whether the signal is rising from a low value or falling from a high one (Figure 4.7). Despite the etymology of the word (ὑστέρησις, meaning 'late-coming'), hysteresis is very important because it **allows cells to make decisions** when presented with noisy inputs. If a cellular response, for example, to differentiate into a new cell type, showed no hysteresis and transitioned between 'stop' and 'go' at a precise level of input, say, 1.0 units, then a noisy input signal fluctuating randomly between 0.95 and 1.05 units would cause the response to switch repeatedly, and stop the cell making a clear decision. If, on the other hand, the system showed hysteresis, switching to 'go' at 1.0 units and remaining at 'go' unless the signal fell below 0.5 units, **small variations would not matter** and the cell will be capable of making a decision and sticking to it unless the input signal suffered a massive fall.

Fig. 4.7 A graph showing hysteresis, in which a system's response to a stimulus follows a different path for low-to-high transitions than for high-to-low. This system requires a high level of signal to switch on, but a lower one to switch off again, so once it has switched on, it will remain firmly on as long as the concentration remains in the blue zone.

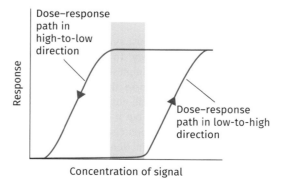

In conventional engineering, for example, electronics, hysteresis is achieved by the use of **positive feedback**, in which an effect of a signal activating a system increases the sensitivity of that system to the signal. Systems biologists examining natural genetic systems thought that they had spotted functional motifs within the networks that would be predicted to have exactly that behaviour. To test the idea, Kramer and Fussenegger constructed the genetic system shown in Figure 4.8. At its core (shown in black in the diagram) is a reporter gene (a gene producing something easy to detect as the output) that is located downstream of a modest promoter, which is provided with binding sites for two extra DNA binding proteins. With neither of these present, the gene would be transcribed at a moderate rate. Also present in the system is a gene encoding a transcriptional repressor (shown in red), which can bind to one of the sites in the promoter of the reporter gene, and hold that reporter gene firmly off. The repressor, though, can be inhibited by the presence of a drug (shown in green). Thus the system as described so far acts as a drug detector: with no drug present, transcription of the reporter gene would be held off but, when the drug is present, the gene would be on, to an extent that follows the concentration of the drug. This system as described so far shows no hysteresis and reducing the concentration of drug would cause transcription rate to decline along the same dose–response line it followed when the drug concentration was being increased.

Hysteresis was added using the extra genetic material shown in blue. This coupled the production of a second gene, encoding a synthetic transcription factor, to the reporter gene. The coupling was achieved by using an **internal ribosome entry site (IRES)**, a sequence of bases in the transcript that allows some protein synthesis to begin at the IRES, and thus translate the second protein-coding sequence in the mRNA, as well as for some protein synthesis to begin at the conventional location near the beginning of the message. IRES sequences are much used in synthetic biology to **produce more than one protein from one mRNA**. The transcription factor encoded by the gene downstream of the IRES binds the other transcription factor binding site in the promoter of

Fig. 4.8 The hysteresis module of Kramer and Fussenegger, made in mammalian cells. The black, red, and green components would generate simple, drug-controlled expression of the receptor. The addition of the blue component, which increases the activity of the promoter of the reporter gene, adds hysteresis, as described in the main text.

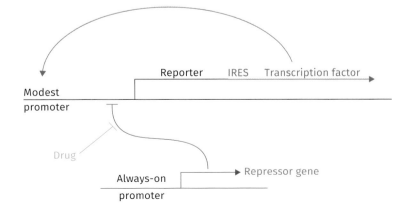

the reporter gene, and thus increases the level of transcription of the reporter and of itself. Thus, once there is enough drug present to activate transcription of the reporter–transcription factor gene, the products of this transcription increase the transcription and the **system drives itself further into activity**. If the concentration of the drug is gradually reduced, the increased sensitivity of the system to activation will ensure that it **remains on** even at levels of drug **insufficient to switch it on**. Eventually, though, when the concentration of drug has been reduced enough, sufficient transcriptional repressor will be active to shut all transcription of the reporter down, and the system will go to its off state. On a graph of activity versus drug dose, the system **followed a different route** in response to a falling drug concentration than it did in response to a rising one, as already depicted in Figure 4.6. This is the hallmark of hysteresis.

Key point

Constructing and testing synthetic versions of network motifs can verify that they do have the expected properties, but this does not itself prove that natural genetic systems are constructed of independent modules or motifs.

Triggering phagocytosis

Phagocytosis is a natural process by which one cell can **engulf pathogens, or the remains of dead cells**, to digest and recycle their material. Analysis of phagocytosis has revealed a very large network of interacting proteins but has suggested a relatively simple theory. In that theory, all that is needed to trigger phagocytosis of dead cells, even in a cell that is not specialized for the behaviour, are two things. One is any protein in the cell's plasma membrane (the membrane that is the cell boundary) that has an **extracellular domain that will recognize dead cells**. The other is the presence of active polymerization of the cytoskeletal protein actin under the cell membrane.

Hiroki Onuma and colleagues used synthetic biology to test this idea. They knew that healthy cells do not show any phosphatidyl serine on the outer surface of their membranes, but dead cells do. They also knew that the **C2 domain** of a protein called MFG-E8 can bind to phosphatidyl serine. They therefore constructed a gene that would encode a novel chimaeric protein formed of parts of three natural proteins. One was the C2 domain of MFG-8, another was the transmembrane domain of giantin, a protein normally found in the membrane of the Golgi apparatus, and the third was cytoplasmic domain of the protein FKBP. Left to itself, this chimaeric protein was produced by the cell and targeted to the Golgi apparatus of a human HeLa cell line, where it stayed (Figure 4.9a). Onuma's construct also contained a gene encoding a second chimaeric protein, consisting of a transmembrane domain from a plasma membrane protein, connected at its cytoplasmic end to an **FRB** protein. This was produced by the cell and inserted into the plasma membrane. The point of producing these rather strange proteins was that the **drug rapamycin causes FKBP and FRB domains to bind together**, and this has the effect of **pulling the C2–giantin–FKBP protein out of the Golgi apparatus and into the plasma membrane**. Onuma and colleagues had therefore built a system that could target C2–giantin–FKBP to the plasma membrane 'on command' when they added rapamycin (Figure 4.9b).

Fig. 4.9 The synthetic cell recognition of Onuma and colleagues. (a) In the absence of rapamycin, each of the synthetic chimaeric proteins is made; the FRB-containing one passes through the Golgi apparatus and travels via the excretory pathway (EP to the plasma membrane), but the C2–giantin–FKBP one remains in the Golgi apparatus. (b) Rapamycin cross-links these proteins and targets the C2–giantin–FKBP protein to the plasma membrane, where the C2 domain can recognize the phosphatidyl serine (PS) residues displayed on the outside of a dying cell.

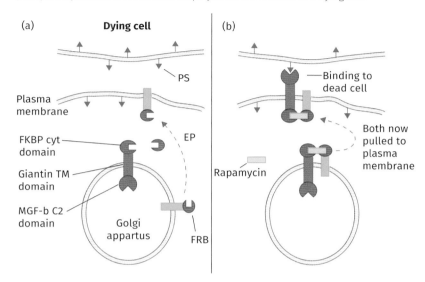

The experimenters then added dying cells to the cultured HeLa cells carrying their synthetic biological device. In the absence of rapamycin, the HeLa cells ignored the dying cells and did not bind to them. In the presence of rapamycin, when the C2–giantin–FKBP moved to the plasma membrane, the HeLa cells **bound to dying cells but still did not engulf them.**

Next, Onuma and colleagues added an extra component, a mutant form of the small GTPase enzyme **Ras** that has been known for a long time to encourage the polymerization of actin. Without rapamycin, the HeLa cells carrying the mutant Ras still ignored the dead cells. With rapamycin, though, they bound them and engulfed them, showing **strong phagocytic behaviour.** This simple system (simple compared to all of the components that are involved in triggering phagocytosis in a real cell) was enough to trigger phagocytosis of dead cells, in HeLa cells that do not normally do it. The technique also worked in other cells that do not normally phagocytose dead cells. It cut through the complexity of normal cellular systems to prove the theory, derived by analytic biologists, that binding of dead cells and active actin polymerization are the two things that together are enough to trigger phagocytosis.

Making patterns

Mammals have **patterned** bodies. Some of these patterns, such as the stripes of zebras or tigers, or the patches of giraffes, leopards, or Frisian cows, are completely obvious, even from the outside. More subtle but still visible externally are fine-scale surface patterns, such as human fingerprints. Inside the skin, the mammalian body is very richly patterned: the repeated 'rib, space' pattern of the thoracic skeleton, the tree-like pattern of airways in the lung, and the forest of villi in the intestine are just three examples drawn from thousands. Viewed in this general sense, including not just patterns of colour but of cell type and of form, **patterning is critical to our development**. It is also scientifically intriguing, because the formation of a pattern where there was none before represents **a gain of order**, and the creation of new, biologically relevant information.

The first serious attempt to understand how patterns might arise in a previously unpatterned system was that of the mathematician Alan Turing, who demonstrated mathematically how a simple network of two interacting molecules that controlled one another's synthesis could generate spots or stripes. His demonstration of this was, incidentally, the first example of computer modelling of a biological process (Turing was also a pioneer of computing). Analytic biologists think they have spotted Turing-type patterning systems at work in a number of embryonic systems and some synthetic biologists are now attempting to build synthetic Turing patterning systems in mammalian cells.

Turing patterning is not, however, the only conceivable way in which patch patterns could arise. Elise Cachat and the author of this chapter proposed that patches could also be produced by a process called constrained **phase separation**. Phase separation is the technical name for the familiar phenomenon by which oil separates from water (Figure 4.10). In this case, the free energy of water molecules is lower when they interact with other water molecules than when they interact with oil molecules. Since physical systems will move to states of lower free energy when they are able to, mixtures of oil and water tend to separate out into a water layer and an oil layer. Mammalian cells stick to one another with adhesion molecules of different types. Some of these show the property that cells displaying the same type of adhesion molecule on their surfaces stick to one another well (meaning low free energy) but cells displaying different adhesion molecules stick to one another much less well (higher free energy). It has been known since the 1960s that, if two populations of cells expressing different adhesion molecules are mixed in a small, three-dimensional aggregate, they will undergo phase separation to make two distinct layers, rather like oil and water do.

We reasoned that, while small, three-dimensional aggregates will allow enough cell movement for complete phase separation to take place, two-dimensional or very large aggregates could not. To understand why, imagine a large, flat, random mixture of cells that have just been commanded to express either of two types of cell adhesion molecule (each cell making its own independent choice).

Fig. 4.10 Phase separation in an unconstrained system, a beaker containing oil and water mixed together by shaking, and then left.

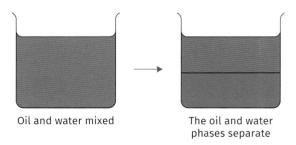

Oil and water mixed   The oil and water
                             phases separate

Once the adhesion molecules are expressed, the cells will begin phase separation, those with the same type of molecule sticking together and excluding those with the other type. The effect of this sticking and exclusion is that patches of cells making one type of cell adhesion molecule will be separated from one another by patches making the other type of adhesion molecule. This has lowered the free energy from that of the random mix, but the free energy could in principle be lower if all cells of one type could come together in one place, minimizing contacts between different types. The problem for the cells, though, is that two patches of type 1 cells separated by type 2 cells could only coalesce if the type 1 cells entered and crossed the type 2 patch, and that would mean a temporary raising of the free energy. This is energetically unfavourable, so does not happen, and the system gets trapped into a **local energy minimum**, unable to reach the state that would give its lowest possible free energy (see Scientific approach 4.1). The same thing would be expected to happen for large, three-dimensional aggregates. Cachat and colleagues built a system to operate this way, in a human cell line, and saw the predicted patterns of patches and stripes (Figure 4.11).

Patterning by phase separation is not something that turns up in textbooks of developmental biology and may be used little, if at all, in natural development. Building a patterning system that uses it is therefore a step in exploring 'roads not taken' by evolution—other ways that something might have been done, but was not (as far as we know). Building mechanisms that did not evolve and comparing them with mechanisms that did evolve may shed light on whether the natural mechanism is better, whether the choice was apparently random, or either would have worked as well. It may also give synthetic biologists novel ways of building tissues, for example, for medical purposes (see Chapter 6).

Fig. 4.11 Pattern formation by phase separation, in a synthetic biological system. The red and green cells can be induced, with a drug, to express different adhesion molecules. Beginning in a random mix they will, when induced, separate partially to produce a pattern of patches.

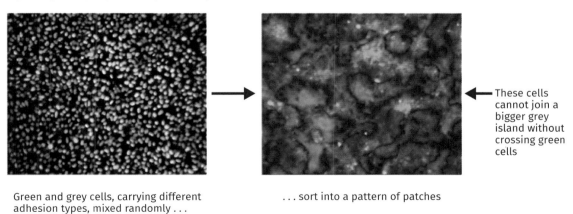

These cells cannot join a bigger grey island without crossing green cells

Green and grey cells, carrying different adhesion types, mixed randomly . . .

. . . sort into a pattern of patches

The patterns shown here come from real data images (with colours intensified for ease of viewing in this book), produced in the author's lab by Elise Cachat.

Scientific approach 4.1
Energy landscapes

Physicists often represent the free energy of a system as a multidimensional 'landscape', in which the free energy at any point is represented by the height of a surface. The idea is illustrated in Figure A, which depicts some measure of system state (e.g. the position of a cell) on the x axis, and the corresponding free energy on the y axis.

Given that systems tend to minimize free energy when this is possible, a system starting on an energy peak will tend to move to states in which the free energy is lower. An example would be water rolling downhill in a literal landscape in a gravitational field. In abstract systems, as with the water, the system can be trapped in a local minimum because it cannot 'borrow' enough energy to escape over a peak to a system that would have the lowest energy.

Fig. A An energy landscape, showing peaks and both local and global minima. Systems tend to flow 'downhill' (minimizing free energy) but can become trapped in local minima.

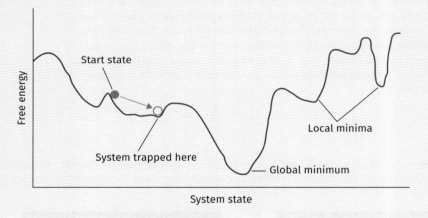

Chapter summary

- Bursts of new discovery in the biological sciences are often triggered by the invention of new research tools: synthetic biology is now creating a range of new tools that allow researchers to work in new ways and to make new discoveries.
- There is potential for a virtuous cycle, in which synthetic biological tools lead to new discoveries about biology, which can be incorporated into better tools, and so on.
- There are two main ways that synthetic biology is used for basic science: (1) as a tool for discovery; and (2) as a way to test the soundness of abstract principles drawn from the analysis of complex natural systems.

- Synthetic biology can be used to add light-sensitivity to neurons, allowing specific brain pathways to be activated in response to targeted laser light.
- RASSLs can be used to identify which pathways in a complicated, multi-pathway response drive particular cellular responses.
- Analytic biologists studying genetic systems believe they have identified 'motifs' that have specific functions and appear in many pathways: synthetic biologists are constructing artificial versions of these motifs to test whether they really do function as expected.
- Analytic biologists studying the complicated pathways of phagocytosis have proposed that all that is really needed is recognition of a target cell and actin polymerization. Synthetic biologists have verified this by making artificial systems that achieve these two things, and showing that they confer phagocytic activity on cells that do not normally have it.
- Synthetic biologists can invent solutions to biological problems, such as pattern formation in an initially unpatterned population of cells, that were not those adopted by natural evolution, and build artificial genetic systems to compare the human-designed systems to the ones that evolved naturally.

 ## Further reading

Davies J (2017). **Using synthetic biology to explore principles of development.** *Development* 144, 1146–58.

Provides an overview of application of synthetic biology to developmental biology and embryology.

Lienert F, Lohmueller JJ, Garg A, Silver PA (2014). **Synthetic biology in mammalian cells: next generation research tools and therapeutics.** *Nat Rev Mol Cell Biol* 15, 95–107.

This article has excellent illustrations.

Mathur M, Xiang JS, Smolke CD (2017). **Mammalian synthetic biology for studying the cell.** *J Cell Biol* 216, 73–82.

Provides an overview of the application of synthetic biology to cell biology.

Wang LZ, Wu F, Flores K, Lai YC, Wang X (2016). **Build to understand: synthetic approaches to biology.** *Integr Biol (Camb)* 8, 394–408.

Provides an overview of synthetic biology as a research tool.

 ## Discussion questions

4.1 What other examples are there of science creating technology creating new science, in a repeated cycle?

4.2 Considering the implications of the addiction experiment in this chapter, should we view addiction as a moral or a medical issue?

4.3 This chapter considered a design of a synbio module that exhibited hysteresis. Using similar components, design one that will remember which of two transcriptional repressor-blocking drugs a cell last experienced, or one that would clamp the transcription rate of an output gene at a set level.

5 TEACHING MAMMALIAN CELLS TO MAKE NEW, USEFUL THINGS

Jamie Billington, Anna Mastela, and Professor Susan J. Rosser

Learning Objectives

- Explain why some medicines and compounds are produced in mammalian cells rather than in bacteria or yeasts.

- List examples of biologics that can be produced in mammalian cells and list challenges associated with producing them that are unique to mammalian cells.

- Describe some of the most commonly used mammalian cell types for the bioproduction of proteins and viruses. Explain why they are well suited to their job.

- Sketch a workflow for how you would go about generating a recombinant CHO cell line.

- Compare how biopharmaceutical companies engineer a mammalian cell line with how a synthetic biologist might engineer one.

- Explain how smart cells implanted into the body might be able to help treat diseases and give examples.

- List some advantages and drawbacks of producing products in whole animals rather than in cell culture.

- Outline how gene-editing technologies are affecting agriculture and the development of livestock.

When attempting to produce biomolecules for industry and medicine, synthetic biologists often to turn to bacterial or yeast cells. Compared to mammalian cells, these microbial hosts grow quickly, are simple to handle, and simple to engineer. A graphical comparison between mammalian and microbial hosts was made in Figure 2.1 in Chapter 2. In many cases though, mammalian cells are our best or only option. It turns out that protein medicines produced by mammalian cell lines are often more effective and safer than alternatives made in microbes. This is due to the post-translational modifications mammalian cells, but not bacterial cells, will make to proteins. These changes can affect the immunogenicity and activity of the proteins in our bodies. Many of the viruses that we use as vaccines, or as vectors for gene therapy, must be produced in mammalian cells if they are to function effectively.

Engineering mammalian cells to manufacture useful molecules can be a big challenge due to the complexity of mammalian gene expression and cell organization (see Chapter 1). In this chapter, we discuss some of the reasons that mammalian cells are used for bioproduction in the first place, the products that they make, and how the challenge of engineering them is approached.

Therapeutic proteins: the next generation of drugs

In recent years, there has been a significant rise in the number of large-molecule medicines that are produced in living cells: medicines produced this way are called biologics. Biologics are often safer and more effective than **small-molecule drugs**, and can be used to treat diseases for which no alternative treatments are currently available. They can be made of proteins, nucleic acids, and sugars (as well as combinations of these entities such as glycoproteins). The majority of biotherapeutics on the market are protein-based, so we will focus on these in this chapter. It worth noting, though, that even therapeutic cells and tissues are considered biologics.

Biologics cannot be synthesized chemically because they are too complex and large. Additionally, they are often heat sensitive and prone to damage by traditional chemical-handling processes. Therefore, biologics can only be manufactured in aseptic, cell-based systems such as microbial or mammalian cell culture. Examples of biologics that are produced in mammalian cells include vaccines, human growth hormones, enzymes to clear blood clots, and antibodies for treating cancers.

Globally the market for biologics is large; it has been valued at £122 billion (2018) and more and more therapeutic proteins are entering the drug development pipeline. Biologists are fuelling this by designing proteins with improved properties over their natural counterparts. Synthetic 'bispecific' antibodies are a good example of this. By recognizing multiple target epitopes, bispecific antibodies can better target cancers with different cell surface markers. Importantly, many emerging biologics derive from mammalian proteins. This means they have evolved to undergo processing by, and therefore to become mature through, mammalian protein expression machinery.

The big challenge associated with developing biologics is that getting from the lab to the patient is slow and incredibly expensive. It can cost billions of pounds and take between ten and fifteen years to bring a new drug to market. Many promising drug candidates therefore fall by the wayside during development.

The vast majority of the expense and time spent reaching the market comes during a research and development (R&D) phase. Each drug must then also pass through three phases of clinical trials to determine its safety and efficacy for use in humans. Conducting these trials is also a costly process, often totalling tens of millions of pounds. These costs are similar for all types of drugs, from small molecules to biologics, but at least they have to be met only at the beginning of a drug's introduction. Biologics are also costly to manufacture at large scale and, because of this, they are expensive and are often used as drugs of last resort. Mammalian synthetic biologists hope to help resolve this by speeding up our ability to do R&D and to scale up biologic production economically.

Biosimilars—biologics for everyone?

One way that biologics can become more affordable is through competition in the drug market. For small-molecule drugs, when a patent covering a popular drug expires, rival companies can release **generics** that are chemically identical to it. A similar scenario occurs in the biopharmaceutical market, where alternative proteins with the same molecular target are released upon a patent's expiry. Unlike generics, these biosimilars are not identical to the drug that inspired them but, as the name implies, they are highly similar. They might have the same protein sequence but their production in different cell lines, by different companies, can affect the complex post-translational modifications they receive. Biosimilars can usually benefit from a streamlined testing process, much less expensive than the trials used for the original biologic (see Case study 5.1). This can significantly decrease the cost and time that it takes to bring a biosimilar to market.

Case study 5.1
Adalimumab and the treatment of inflammation

The biggest-selling drug on the pharmaceutical market from 2012 to 2016 was a therapeutic protein called adalimumab. Adalimumab is an anti-inflammatory medicine, produced in mammalian cells, which is used to treat diseases such as rheumatoid arthritis and Crohn's disease. The protein is an example of a **monoclonal antibody (MAb)**: in this case it binds a protein called tumour necrosis factor alpha (TNF-alpha) in the blood. TNF-alpha is responsible for stimulating an overactive immune response, which underpins inflammatory diseases like arthritis. By neutralizing and removing TNF-alpha from the circulating blood, adalimumab can help to alleviate inflammation.

Global sales of adalimumab exceeded $16 billion. It was knocked from the top of the list of biggest-grossing drugs only when the patent covering it expired in the USA. Other companies were therefore able to release their own biosimilars—alternative MAbs that also targeted TNF-alpha. Prior to its patent expiring, adalimumab was being sold for approximately $4370 per month. Now these competitor drugs cost as little as $200 per month, making them more accessible to patients.

Producing viruses

As well as protein production, the other major use of mammalian cells in industry is producing viruses. Viruses are used in medicine, either as **vectors** for delivering DNA into cells, or as **vaccines** to immunize people against pathogenic viruses. Most of the viruses that we are interested in are structurally simple, can self-assemble, and have evolved to reproduce in mammalian cells. Mammalian cells can therefore be engineered to produce viruses by transfecting the cells with DNA or RNA that encode the viruses' component parts (see Case study 5.2).

Case study 5.2

Virus production within mammalian cells is becoming increasing important for generating vaccines against the **influenza virus**. Influenza is a rapidly evolving virus that frequently changes its surface **antigens**, thus avoiding detection by immune systems that have learned to detect previous versions of the virus. Developing an effective vaccine every year against emerging strains has therefore always been a major challenge.

Historically, the year's influenza vaccine was produced by expanding an attenuated form of the virus inside chicken eggs. This was an inconvenient way to produce large amounts of vaccine as it was slow and it could take one to two eggs to produce one dose of vaccine. With no alterna-tive, any disturbance in the egg-vaccine supply chain could have resulted in our inability to prevent a pandemic. The first alternative vaccine produced in mammalian cells finally became available in 2012. Flucelvax, produced in a canine cell line called Madin–Darby kidney cells, can be made with a much quicker turnaround than vaccines made in eggs.

Synthetic biology is helping to further speed up the vaccine development cycle through the synthesis and assembly of viral genomes. In a project headed by Novartis, researchers built **hybrid viral genomes** com-posed of constant regions that are common to influenza strains, and chemically synthesized variable regions. These variable regions represent the viruses' unique sur-face antigens and were designed based on sequencing data obtained from an influenza outbreak site. By taking this approach, the team was able to go from sequencing data to a live vaccine growing in cells within twelve days.

Obviously, producing large quantities of viruses that can infect humans poses a major risk to the people doing the work. To mitigate this risk, scientists gener-ally work with **replication-incompetent** versions of viruses that cannot repro-duce in normal human cells. These viruses can be produced only in mammalian cell lines that have been engineered to express viral replication machinery missing from the genes of the incompetent virus itself. **RNA-based viral vec-tors** that cannot integrate into our genomes are also preferred to DNA-based viral vectors (which might), for similar reasons.

Some of the viruses produced industrially include the poliovirus and influ-enza virus for vaccines, and adeno-associated virus (AAV), which is used in gene therapies. **Gene therapies** are an emerging method for treating genetic diseases. They involve delivering functional copies of a gene into a patient's cells to perform the function that is missing in someone with a defective gene. This opens up ways to cure previously untreatable diseases such as retinitis pigmentosa, an inherited form of blindness that can arise as a result of a single mutation in any of fifty genes associated with the development and function of the retina. Viruses are ideal carriers for therapeutic genes due to their ability to target specific cell types, to deliver DNA across the cell membrane, and to evade destruction by our immune systems.

 Key point

Mammalian viruses have applications in medicine for producing vaccines and as vectors for delivering DNA into mammalian cells. Synthetic biology can be used to engineer both viruses and the cells that they are produced in to make more effective and safer medicines.

Cell factories for protein production

Protein-based biologics are produced in many organisms including bacteria, algae, yeast, and immortal cell lines from insects, plants, and mammals. These systems all have their own advantages and disadvantages, which are listed in Table 5.1. Usually choosing between them depends on the nature of the protein, the application, and how much of it we need to make. In the vast majority of cases though, biologics are manufactured in mammalian cells.

Table 5.1 Comparison of different expression systems for production of biologics

Type	Expression system	Advantages	Disadvantages	Class of biologics produced	Example available on the market
Unicellular system	**Bacterial**	Can be scaled up Low cost Chemically defined media available Quick growth High yields of product Easily engineered	Many PTMs not available (lack of appropriate enzymatic machinery) Recombinant proteins may aggregate inside of the cell (lack of chaperons) May require protein-specific optimization Must be extensively tested for **endotoxin** presence	**Antibody formats** Fc-fusion proteins Cytokines Enzymes Peptides Therapeutic toxins	**Cimzia®** (certolizumab pegol)—TNF-alpha inhibitor (recombinant humanized MAb binding fragment covalently bond to ethylene glycol)
	Algal	Robust selection and expression Low cost Can fold proteins correctly High yields	Lack of glycosylation Technology still in development	May be suitable for production of complex **glycoproteins**, i.e. MAbs	No therapeutics available on the market (statement correct in early 2018)
	Yeast	Easily scalable Low cost Chemically defined media available Quick growth High yields of product Eukaryotic processing of proteins Easily engineered Considered to be GRAS (generally recognized as safe)—no need for extensive safety tests	Glycosylation profile different from mammalian For high yields, large-scale fermentation is required Low efficiency of secretion	Enzymes Peptides Clotting factors Vaccines	**Gardasil®9** (recombinant vaccine against human papillomavirus 9-valent) **Victoza®** (liraglutide)—a recombinant glucagon-like peptide-1 receptor agonist for the treatment of type 2 diabetes

	System	Advantages	Disadvantages	Products	Examples
	Insect	Mammalian-like PTMs available; Higher yields compared to mammalian systems	Demanding culture conditions; Time-consuming; Glycosylation profile different from mammalian	Vaccines	**Cervarix®** (a vaccine against human papillomavirus) produced in baculovirus
	Mammalian	Human or human-like PTMs available; Correct assembly and folding of complex proteins; Secretion of recombinant proteins is feasible; Proteins can be expressed either transiently or stably	Demanding culture conditions; High yields achievable only in large-scale suspension cultures; Expensive media; Lengthy and laborious cell line development process; Genomic instability	MAbs; Cytokines; Enzymes; Fc-fusion proteins; Hormones; Clotting factors	**Herceptin®** (trastuzumab)—a MAb against human epidermal growth factor receptor 2 (HER2); used to treat certain types of breast cancer
Multicellular system	**Transgenic plants**	Low cost; Can fold proteins correctly; Can be engineered to secret products; High yields	Low transformation efficiency; Glycosylation profile different from mammalian	Enzymes; Prospect of manufacturing **edible vaccines** and MAbs	**Elelyso®** (taliglucerase alfa)—a recombinant **glucocerebrosidase (a house-keeping enzyme)** produced in recombinant transgenic carrots
	Transgenic animals	Human or human-like PTMs available; Correct assembly and folding of complex proteins; Secretion of recombinant proteins into milk is feasible	Time-consuming production; High cost; Low yields	Clotting factors	**ATryn®** (recombinant humanized antithrombin III, an anti-coagulation factor) produced in a recombinant transgenic goat
	Cell free	Open system (unnatural additives can be easily added); Quick expression; Easier **downstream processing** of products; Ability to perform correct PTMs and folding	Production on large scale is currently unavailable; Technology still in development	MAbs (reported); Technically can be used to express any kind of protein; Especially suitable for production of toxic proteins	No therapeutics available on the market (statement correct in early 2018)

Obtaining the right decorations

Like the native proteins in a cell, a therapeutic protein requires post-translational modification (PTM) to function. The variety of possible PTMs that can be added to a protein is staggering. Lipidation, phosphorylation, acetylation, methylation, and ubiquitination are just a handful of the myriad possible modifications a protein can receive. Each modification requires a dedicated set of enzymes and can affect the folding, enzymatic activity, and stability of a protein. Ultimately, these factors will determine the safety and efficacy of the protein as a drug.

One variety of PTM that is particularly important from a biomedical standpoint is glycosylation. In this process, oligosaccharides (short, and sometimes branched, chains of sugars) are built on to a protein's amino acids by enzymes known as **gylcosyl-transferases**. Different organisms contain unique combinations of glycosyl-transferases so the type of glycans a protein receives will depend on the organism in which it is produced. Yeast cells produce side chains that are rich in the sugar **mannose** and mammalian cells produce side chains that almost entirely lack mannose.

The biggest advantage that mammalian cells have over their competitors is their ability to perform **human-like PTMs** and **correctly fold/assemble** complex, multi-domain proteins. This is thanks to the protein-processing machinery that they contain. The correct post-translational processing of proteins affects their ability to achieve full biological activity as well as their serum half-life, meaning how quickly they will be removed from the bloodstream of patients.

An unusual feature of human glycoproteins is that our cells lack a sugar building block called *N*-**glycolylneuraminic acid** (Neu5Gc) due to a mutation in the enzyme that synthesizes it, a mutation that occurred sometime during the evolution of humans from ape-like ancestors (modern apes still make Neu5Gc). Our immune systems have evolved to exploit this fact, recognizing glycans that contain non-human Neu5Gc. Proteins with Neu5Gc are targeted for destruction and have a lower serum half-life. By producing therapeutic proteins in human cells, or other mammalian cells in which the gene producing Neu5Gc has been removed, we can ensure the oligosaccharide side chains that a protein receives are as similar as possible to those found in humans and unlikely to induce an immunogenic response.

 Key point

Mammalian cells are used to produce therapeutic proteins because they are more likely to perform the correct combination of post-translational modifications for the protein to function effectively in humans. This is especially true when the mammalian cell is derived from a human.

Mammalian cell lines
CHO cells for bioproduction

For the production of therapeutic proteins, by far the most commonly used cell factories are **Chinese hamster ovary** (CHO) cells. Seventy per cent of all recombinant proteins on the market are made in them, including therapeutic MAbs such as rituximab and the first-ever US Food and Drug Agency (FDA)-approved recombinant protein, tissue plasminogen activator. CHO cells have risen to this dominant position due to several favourable characteristics. The most important of these is that CHO culture can be easily scaled up for growth in giant industrial bioreactors.

The other big advantages of CHO cells are that they can add many (but not all) **human-like** PTMs to proteins, they effectively fold and secrete them, they can grow in chemically defined, serum-free media, and they can produce large amounts of protein in optimized conditions. Growth in chemically defined media is important because it means we can ensure that there is no variation between batches of media. This reduces the variability in yields between production runs. From a biosafety perspective, CHO cells are also favourable because they cannot transmit human viruses that could be passed on to patients.

Despite these advantages, CHO cells are far from an ideal expression platform—unfortunately, their glycosylation profile is not an exact match to the human one. CHO cells produce some types of glycans with sugars that can trigger an immunogenic reaction in patients (i.e. Neu5Gc) and they cannot perform all varieties of human glycosylation such as α-2,6-sialylation and/or α-1,3/4-fucosylation. This requires the manufacturer to apply an extensive screening process to find a CHO cell line that will have a correct glycosylation profile. This increases the length and cost of the cell line development process. Also, the genome of CHO cells is innately unstable, undergoing chromosomal rearrangements over the course of a production run. This can affect how much protein each cell produces over time and result in a gradual decline in cell line productivity. Fortunately, synthetic biology can offer a solution to these problems (see 'Using synthetic biology to improve cells for bioproduction').

Human cell lines

Human cell lines are popular alternatives to CHO cells due to the non-immunogenic, fully human PTMs they add to proteins. The list of industrially relevant human cell lines includes human embryonic kidney 293 (HEK293) cells, HT-1080 (derived from fibrosarcoma), and Per.C6 (from embryonic retinoblasts). Dulaglutide, a protein therapeutic for treating type 2 diabetes mellitus, is one of many examples of a protein that is produced in human cells.

Like CHO cells, some human cell lines can also be grown in large-scale suspension cultures with chemically defined media and reach high production yields. A major concern in using them, though, is that viral infections in these cells might be transferred to patients along with the drugs we extract from them. Regulatory agencies also prefer CHO cell lines as there is more clinical data available for products made in them. This helps to speed up the approval process for new medicines.

Protein production as an industrial process

The goal of industrial production of biologics is to produce the highest quantity of a protein of the best possible quality. This requires biotech companies to find a balance between exploiting the cells' machinery to maximize synthesis of the required protein and ensuring that cells can grow healthily at the same time. Also, due to the biosafety regulations, the whole process has to be rigorously controlled and kept in line with **current good manufacturing practice**. Generally, while transient expression can be used to generate small batches of drugs quickly, for example, for clinical studies, all biologics on the market are made in stable cell lines.

Cell line development

As the CHO cells are the most commonly used cell type for manufacturing of biologics, this section discusses the typical industrial process of developing a production CHO cell line.

Case study 5.3
A Lesson learned from cancer cells

In the 1950s, the Hakala group discovered that the cancer cell line that they were working with had developed a resistance against an anti-cancer drug called **methotrexate** (MTX). MTX works by specifically blocking the activity of **dihydrofolate reductase (DHFR)**, an enzyme involved in the synthesis of purines, some amino acids, and thymidylic acid. Consequently, inhibition of DHFR impairs cell proliferation and growth. The initial data that they obtained showed that the amount of DHFR enzyme in this MTX-resistant cell line had somehow become elevated, allowing some enzyme activity to be present even in the presence of MTX. Further investigation by different research teams revealed that this elevation was due to an increase in the number of copies of the *dhfr* gene. This turned out to be a consequence of **gene amplification** in response to high doses of MTX and/or repeated administrations of this drug.

The potential of this discovery was quickly recognized by the biopharmaceutical industry, who turned it into an alternative selection technology that replaced antibiotic resistance-based selection. Antibiotic selection is in many ways problematic in industry due to the relatively high costs and potential safety implications. This was therefore a big step forward in cell line development enabling the production of large amounts of proteins in a safe manner.

The first step in the cell line development process based on MTX selection (see Case study 5.3) is to **transfect** host cells with a suitable expression plasmid. A typical vector includes a promoter, a polyA sequence, a selection cassette (in this case a *dhfr* gene), and, of course, a transcription unit encoding a gene of interest (GOI) (Figure 5.1a).

Typically, DNA is delivered into a host cell line that lacks the DHFR enzyme as a linearized plasmid by electroporation or nucleofection. These are fast, efficient, and cost-effective ways of getting large segments of DNA into the cells (Figure 5.1b). Auxotrophic CHO cell lines that have had the *dhfr* gene knocked−out (CHO-DG44 and CHO-DXB1) are common host cell lines for integration. Prior to transfection, these cells are cultured in a media supplemented with hypoxanthine and thymidine, chemical intermediates that can complement the absence of DHFR. After transfection, once they have taken up DNA, they are moved to a selective medium that lacks hypoxanthine and thymidine and, thus, prevents the growth of non-transfected cells. Only the cells that **stably integrate** the DNA we provide into their genome will continue to grow.

New **clonal** cell populations are then grown from single cells and subjected to multiple rounds of MTX-induced **gene amplification**. Increasing selection pressure (achieved by step-wise addition of MTX to the media cells are grown in) can specifically increase the number of *dhfr* gene copies and, because they are closely connected on the same piece of DNA, copies of our GOI. We then determine how much protein different cell clones produce and further **expand** high-producing clones. Finally, the best producing clones are **evaluated** in control bioreactor runs and **banked** for future manufacturing purposes.

All drug regulatory agencies require that the biologics be produced in monoclonal cell lines derived from a single cell with a biologic encoding gene stably integrated in its genome. This helps to ensure that the manufactured product remains unchanged throughout a production process that may continue for years.

Fig. 5.1 The MTX selection system in DHFR-deficient CHO cell line.

(a) An expression vector for the MTX selection system

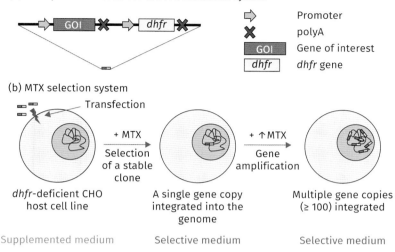

(b) MTX selection system

When the MTX selection system was being developed in the early 1980s, cells that lacked the *dhfr* gene were made using random chemical mutagenesis, followed by gamma irradiation. This resulted in the CHO-DXB11 cell line, a cell type with a single mutant *dhfr* copy. These cells were prone to **spontaneous reversion,** where their remaining functional dhfr copy repaired the mutant version, so lots more work had to be done to generate a double knockout strain. Nowadays, CRISPR/Cas9 genome editing is making knocking out these genes almost trivial. This has exciting potential to increase rapidly a number of alternative hosts/selection systems we can develop and to tailor cell lines to our needs.

 Key point

Steps of **recombinant CHO cell line** development process:

- transfection of host cells with a plasmid DNA containing a GOI and a selection cassette;
- selection of single-cell clones that have stably integrated a GOI into the genome, followed by gene amplification;
- screening for the highest-producing clones and their further expansion;
- evaluation of selected clonal cell lines and cell banking.

The MTX amplification system enables us to both the select cells we are interested in and improve protein production by amplifying genes of interest.

Using synthetic biology to improve cells for bioproduction

It is important to remember that mammalian cells never evolved to work as cellular factories. The industrial process of generating recombinant cell lines therefore often runs into issues, including variability between batches of cells, gene silencing, and difficulties in expressing some proteins. Collectively these issues can slow progress, prove costly, and hinder the development of promising drugs. Synthetic and systems biologists believe that new tools and techniques for engineering biological systems might offer potential solutions to these problems.

Engineering better glycosylation

As previously highlighted, ensuring that therapeutic proteins receive the right set of PTMs is an important aspect of protein manufacturing. To engineer CHO cell lines that produce more human-like glycosylations, synthetic biologists are trying to alter glycosyl-transferase gene expression in CHO cells. One strategy being explored is using CRISPR/Cas9 to delete glycosyl-transferases that work in undesirable glycosylation pathways. Another is to try and repress or activate certain glycosyl-transferases to favour certain reactions in the glycosylation pathway. Synthetic transcription factors based on dCas9, a nuclease-dead form of Cas9 (further discussed in Chapter 6), are a promising tool for attempting this, due to their ability to target multiple genes easily and simultaneously.

Reconfiguring metabolic pathways

Beyond just altering glycosylation patterns, synthetic biologists are engineering mammalian cell metabolism to optimize protein manufacturing. One example of this is the introduction of a pyruvate carboxylase (PYC2) enzyme from yeast into CHO cells. PYC2 processes a toxic waste metabolite called lactate into pyruvate, an important component of central carbon metabolism. It was shown that removing the toxic lactate in this way both improved the health of CHO cells in culture and their ability to produce antibodies.

Targeted integration of genes

Another area where synthetic biology is being applied to improve mammalian bioproduction is to target our genes of interest into **transcriptional hotspots** in the genome. Hotspots are gene-rich locations in the genome that favour high levels of transcription. One way to find these sites is to engineer genes into transposons—'jumping genes' that integrate semi-randomly into the genome. If the transposon carries a landing pad/site-specific recombinase target site, then after the transposon has inserted, it can be targeted for delivery of new genes via recombinase-mediated cassette exchange. This allows good sites to be reused, and importantly allows us to compare the expression of different genes inserted in a particular location. The first few studies utilizing these approaches report that the genes expressing antibodies introduced into CHO cell landing sites in hot spots appear to be stably expressed for longer.

 Key point

Synthetic biology tools such as genome editors, landing pads, and genes can be applied to mammalian cell culture both to improve the health of cells and to create cells that produce higher-quality proteins.

Producing therapeutic proteins in the body

Rather than simply converting mammalian cells into protein factories, synthetic biologists also aim to design 'smart' cells that can be implanted into our bodies and produce biomolecules where, when, and in the concentration they are needed. Cells containing **theranostic** (a combined term for therapeutic and diagnostic) circuits are an example of this kind of 'smart cell', which show how synthetic cells could help treat disease. At their core, smart cells rely on the connection of the output of a *diagnostic* biosensor to a *therapeutic* effector. Cells are full of molecules and systems that can detect and manage disease states so developing theranostic circuits is often a case of reconnecting natural components in new ways. With the correct components, an engineered cell could theoretically monitor and manage a chronic disease all by itself, reducing or even removing the need for medical intervention.

Balancing blood sugar: circuits for treating insulin resistance

One of the most exciting recent examples of theranostic circuits are those that are able to detect and respond to insulin. Insulin, as you might be aware, is a hormone responsible for regulating blood sugar levels. Essentially, it instructs cells, including fat cells, to take up and store glucose from the blood. In people with type 2 diabetes mellitus, fat cells become unresponsive to insulin signalling and fail to take up glucose. This phenomenon, called insulin resistance, leaves the sugar in the blood and can lead to chronically high blood sugar levels. These high sugar levels damage blood vessels and can lead to a host of health complications such as cardiovascular disease, kidney disease, and blindness.

As a proof-of-concept for how to resensitize cells to insulin, a human cell line has been constructed that contains an engineered insulin response, independent of the natural one. Normally, the mitogen-activated protein kinase (MAPK) signalling pathway coordinates the response to insulin by activating the transcription of a handful of genes. This response to MAPK signalling is diminished in insulin-resistant cells. To compensate for diminished signalling, the researchers built a protein biosensor of MAPK activity that triggered the production of a synthetic **adiponectin** gene. Adiponectin is a soluble human protein that spreads through the blood and naturally influences cell sensitivity to insulin. They showed that, in a diabetic mouse, these theranostic cells released adiponectin in an insulin-dependent manner and that this improved the mouse's tolerance to glucose and improved its symptoms related to diabetes.

Gout-be-gone: circuits for removing urate

Another disease that might be amenable for treating with smart cells is gout, a disorder that develops when a person cannot properly break down the chemical urate. Urate is a natural waste product generated by the breakdown of metabolites called purines, which we normally excrete. In certain people who can't effectively clear urate (about one per cent of the Western population), it can accumulate to form crystals in their joints. These crystals can be incredibly painful and gout flare-ups can be a persistent problem throughout a person's life. Simply eliminating urate is not an ideal solution, though, because while high concentrations of urate are dangerous, at normal concentrations the molecule performs a valuable function as an antioxidant, protecting tissues from damage and premature ageing.

The same group of researchers who built the insulin-detecting circuits also engineered cells that manage urate levels. They did this by identifying a urate-sensitive transcription factor called HucR from a bacterium (*Deinococcus radiodurans*). They built a synthetic mammalian promoter that contained the DNA sequence to which HucR naturally binds (hucO), and then showed that this promoter responded to high levels of urate (Figure 5.2). To manage these high

Fig. 5.2 The synthetic biological diabetes-suppressing and the gout-suppressing systems that have been demonstrated in living mice.

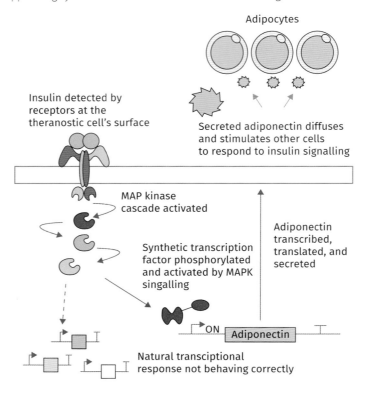

Adipocytes

Insulin detected by receptors at the theranostic cell's surface

Secreted adiponectin diffuses and stimulates other cells to respond to insulin signalling

MAP kinase cascade activated

Synthetic transcription factor phosphorylated and activated by MAPK singalling

Adiponectin transcribed, translated, and secreted

ON Adiponectin

Natural transciptional response not behaving correctly

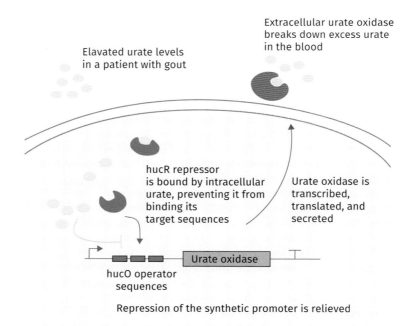

Extracellular urate oxidase breaks down excess urate in the blood

Elavated urate levels in a patient with gout

hucR repressor is bound by intracellular urate, preventing it from binding its target sequences

Urate oxidase is transcribed, translated, and secreted

Urate oxidase

hucO operator sequences

Repression of the synthetic promoter is relieved

urate levels they linked their promoter up to a urate oxidase enzyme from a fungus called *Aspergillus flavus*. When urate was present, these cells produced and secreted urate oxidase, reducing urate levels in the blood of a previously gout-prone strain of mouse that had been given implanted engineered cells. This action helped to alleviate its gout symptoms.

In both of these cases, by putting therapeutic outputs under control of a sensor, synthetic biologists could control the production of 'medicines' within the body to make sure that they were only produced when and where they are needed. This could be important for avoiding any complications that might arise when therapeutic molecules are constantly present or present where they are not needed (potentially causing toxic effects). These gout detectors and insulin detectors illustrate how cells can be made smarter both by the rearranging of natural circuits and by building entirely new ones.

 Key point

One option for treating chronic diseases is to implant cells that produce a therapeutic protein into the body. Theranostic circuits link a diagnostic biosensor component and therapeutic output. These can be introduced into implanted cells to confer the ability to sense a disease state and produce the therapeutic when and where it is required.

Going agricultural: synthetic biology in farm animals
Pharming and medicines from milk

Although cell culture is currently the method of choice for producing therapeutic proteins outside the body, that may not be the case forever. Pharming, a blend of the words *pharmaceuticals* and *farming*, aspires to engineer animals for production of specific proteins. In this field, pigs, goats, and cows are engineered to secrete designer proteins into their blood, milk, or urine through precise control of gene expression. Technologists taking this approach believe that this approach might reduce the costs of producing protein drugs and allow us to scale up their production to truly huge quantities, in a manner independent of a continuing large-scale infrastructure of bioreactors.

Pharming has become feasible in the past thirty years due to major advances in our ability to manipulate the DNA of large animals. Designer DNA can be introduced into an unfertilized or fertilized egg by micro-injection or infection with a retrovirus. In some treated cells, this DNA will randomly integrate into the genome. Researchers implant the cells they have treated into a host mother and identify embryos that have incorporated this DNA into their genome by screening techniques such as polymerase chain reaction (PCR). Once one animal that produces a biomolecule has been generated, researchers clone it or breed from it conventionally, to generate a population of transgenic animals.

To direct the production of the therapeutic protein to a specific organ within an animal, genes encoding the protein of interest are placed under the control of a tissue-specific promoter. Tissue-specific promoters are transcribed only when the correct cocktail of transcription factors is present. The most useful promoters for pharming are those that are active in a lactating mammary gland, such as the prolactin-sensitive promoter (Figure 5.3). Mammary glands produce

Fig. 5.3 Pharming to produce a protein in cows' milk.

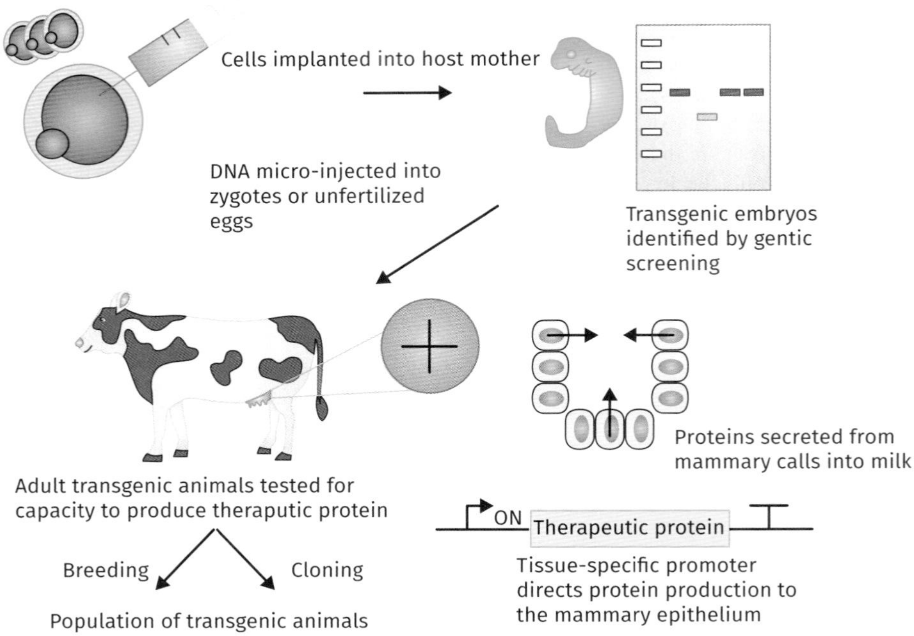

Cells implanted into host mother

DNA micro-injected into zygotes or unfertilized eggs

Transgenic embryos identified by gentic screening

Proteins secreted from mammary calls into milk

Adult transgenic animals tested for capacity to produce theraputic protein

Breeding   Cloning

Population of transgenic animals

ON Therapeutic protein

Tissue-specific promoter directs protein production to the mammary epithelium

large amounts of protein and secrete it in milk. This has the added bonus of making proteins relatively easy to extract using the infrastructure of the dairy industry.

Pharming has already had several successes. For example, an anti-blood-clotting factor made in goats has already been approved for sale in the USA. There is now intense research into the idea of transferring many biologics currently produced in CHO cells in bioreactors to production in living animals. To illustrate the potential advantage of pharming, a gram of therapeutic protein is estimated to cost between £250 and £2500 to produce in mammalian cell culture, but only £80 in a transgenic goat (2018 prices). Scaling up mammalian cell culture requires expensive bioreactors and facilities for growing cells. Scaling up pharming just requires the breeding of more animals.

Improved livestock

Besides the potential for producing new medicines in livestock, mammalian synthetic biology has excited interest in agriculture because of its potential to speed up the improvement of livestock. We have been genetically modifying livestock by selective breeding for thousands of years, but have been limited by the generation times of animals we breed. Genetic modification with synthetic biology could take features that might take many generations to breed in and deliver them in a single step.

'Enviro pigs', a variety of pig engineered to express a plant enzyme called phytase in their salivary glands, are a good example of this. Pigs are fed phytase as a supplement to help them absorb phosphate from feeds like corn and cereal grains. The phosphate that they do not absorb ends up in their waste, which can be a big problem if it gets into the water system. Phosphates fuel the growth of algal blooms in freshwater ecosystems, which absorb oxygen from the water and kill animal and plant life. Pigs that produce phytase for themselves are better able to absorb phosphates, produce less phosphate waste, and reduce the risk of algal blooms being created.

Perhaps the most promising emerging technology for livestock improvement is gene editing. CRISPR/Cas9, the gene editor discussed in Chapter 3, has quickly been adopted by the research community to generate breeds of cattle without horns and pigs with altered fat contents to change the taste of their meat. Recently, CRISPR-edited pigs were generated to be resistant to porcine reproductive and respiratory syndrome. This disease has previously been a major problem for farmers and been estimated to have an economic impact of £420 million per year.

At present, genetic modifications are all relatively simple, often involving single base-pair changes. As gene editing becomes more sophisticated though, it may become possible to transform the shape and functions of animals completely by making multiple edits simultaneously. Big questions hang over the ethics of making these kinds of modifications, particularly relating to how they might affect the well-being of the animals. The public's long-held distrust of genetically modified organisms is unlikely to help the people who argue for this kind of engineering. A more comprehensive look at some of the issues we have touched on here will be discussed in Chapter 7.

 Key point

Mammalian synthetic biology offers new ways to speed up the improvement of livestock and engineer them for the production of new products. Gene editing technologies in particular are likely to play a significant role in the development of agriculture in the next few decades.

 ## Chapter summary

- Mammalian cells are widely used in industry to produce protein-based medicines. This is due to their capacity to perform sophisticated post-translational processing reactions.
- Mammalian cells also are an emerging technology for the large-scale production of viruses for vaccines and gene therapies.
- The market for medicines produced in mammalian cells is growing and companies require advances in cell line development to deliver new products.
- CHO cells are the premier cell line for producing therapeutic proteins.
- CHO cell line development uses chemical selection and gene amplification to produce large amounts of protein.
- Synthetic biology can be applied to mammalian cell line development through metabolic pathway engineering, rewiring gene expression, and the integration of new circuits into cells.

- Smart cells, designed to sense and produce therapeutic proteins in response to disease in the body, are currently under development.
- Pharming, the engineering of livestock to produce therapeutic proteins is a potential cost-effective alternative for scaling up bioproduction.
- Gene editing promises to have a major impact on our ability to engineer animals and improve livestock.

 ## Further reading

Evolution Global (n.d.). The future of biomanufacturing: cell line engineering. https://evolutionexec.com/the-future-of-biomanufacturing/cell-line-engineering/.

Good overview of the companies involved in biopharma, the products being made, and the state of the biopharmaceutical industry.

Genetic Science Learning Center (2013). Pharming for farmaceuticals. http://learn.genetics.utah.edu/content/science/pharming/.

Great interactive figures covering the process of generating transgenic animals.

Lai T, Yang Y, Ng SK (2013). Advances in mammalian cell line development technologies for recombinant protein production. *Pharmaceuticals (Basel)* 6, 579–603.

Helpful explanations and some more detail about CHO cell line development.

Lee JS, Grav LM, Lewis NE, Faustrup Kildegaard H (2015). CRISPR/Cas9-mediated genome engineering of CHO cell factories: application and perspectives. *Biotechnol J* 10, 979–94.

Covers in more depth the idea of using genome editing to improve cells for bioproduction.

Rutherford A (2012). Synthetic biology and the rise of the 'spider-goats'. https://www.theguardian.com/science/2012/jan/14/synthetic-biology-spider-goat-genetics.

Interesting article on using pharmed goats to make spider proteins. Related to an interesting Horizon documentary on synthetic biology.

 ## Discussion questions

5.1 You have been asked to produce a new antibody-based biologic, Novelimab. How would you choose your cell-based production system, which technologies would you use, and what problems do you think you might encounter along the way?

5.2 There is a new respiratory virus sweeping the globe. You and your team have been tasked with producing a vaccine to prevent more people in the UK becoming sick. You have got samples from infected patients; what do you do next? How are you going to develop your vaccine and how are you going to produce it in sufficient quantities to rapidly distribute it around the country?

5.3 You have been given a lab, funding, and a team of scientists for a pharming research project to produce a high-value soluble protein, PhmX. Which animal would you choose to produce it and why? Describe briefly your research programme and how you would obtain your protein. What ethical issues might arise as a result of this work?

6
SYNTHETIC BIOLOGY, STEM CELLS, AND REGENERATIVE MEDICINE

Professor Steven M. Pollard

Learning Objectives

- List the key features of stem cells.
- Sketch the various sources and applications of stem cells.
- Describe the key challenges of differentiating pluripotent stem cells to specific cell and tissue types.
- Explain how synthetic transcription factors can help us control stem cell behaviour.
- Describe the potential applications of mammalian synthetic biology for new types of tissue repair and regeneration.

Our bodies are made up of a huge variety of cell types that carry out the multitude of processes required for our daily functions. Cells operate as integrated specialized communities in the form of tissues and organs. How are the vast varieties of different functional cell types generated during our embryonic development? How are our adult tissues maintained throughout life? How are new cells generated upon injury or disease? These are questions that have been tackled by developmental biologists and stem cell biologists and, over the past decades, they have made remarkable progress in discovering many of the molecular processes involved. Knowledge of stem cell biology is stimulating new kinds of **regenerative medicine**; that is, medicine designed to rebuild damaged or missing tissues. This might be achieved using either laboratory-grown cells or by manipulation of the behaviour of the body's natural **stem cells**.

In this chapter, we discuss how stem cell technologies might connect with and use synthetic biology. This is an incredibly exciting prospect. We discuss the various types of stem cell and how the new and emerging toolbox of synthetic biology is unlocking new ways to control and manipulate mammalian stem cells. These experimental tools also provide new ways to dissect the molecular processes controlling cell form and function. Can we exquisitely control cell behaviours through engineering of new genetic circuits? Can we create new cellular capabilities—not just fixing and re-creating the tissue that was there before—but imparting completely new cellular 'superpowers', to enable better early detection of disease, abolish invading pathogens or malignant cells, and rejuvenate our tissues and organs to restrict the ageing process? New types of stem cell engineering are emerging that can deliver us a variety of new applications. For example, as biosensors or diagnostics (e.g. for early detection of cancer or for drug-screening assays); for engineering complex genetic circuits involved with cell and gene therapy (e.g. controlled release of proteins such as insulin, with feedback circuits); in new types of cell transplantation medicine (e.g. for efficient production of desirable cell types); or for biomanufacturing (e.g. production of molecules that are useful for therapeutics).

What are stem cells?

Stem cells can continuously divide to produce unaltered daughter cells—a process termed self-renewal. But they also have an ability to generate cells with specialized function (e.g. liver cells, neurons, and skin cells) through a process termed differentiation. Stem cells are effectively immortal; they last the lifetime of the organism. However, unlike cell transformation and escape from the Hayflick limit (discussed in Chapter 2), stem cells naturally express the enzyme telomerase without being genetically altered.

It is helpful to think about two major categories of stem cell: embryonic stem cells (ESCs) and tissue stem cells. These have different properties and features that make them well suited to exploring distinct biological questions and use in different applications. A key difference is that ESCs have a potential to make any cell type of the adult organism; they are referred to as **pluripotent**. By contrast, tissue stem cells, whether found in the developing fetus or in adult tissues, have more limited differentiation capacity. They generally only produce differentiated cells of the tissues in which they reside; that is, blood stem cells make blood and brain stem cells make brain cells (neurons and glia). Harnessing the potential of stem cells requires knowledge of how they self-renew and differentiate—both in the laboratory culture dish and in living organisms (*in vivo*). This one of the most exciting and promising areas of biomedical research.

Embryonic stem cells

At the very earliest stages of human development, a single cell embryo undergoes a series of cell divisions, creating a cluster of several hundred cells that will produce all the tissues and organs of the adult human body. These cells— officially termed inner cell mass cells of the blastocyst—are pluripotent, and can be 'captured' in the laboratory culture dish and grow continuously as stem cell lines using a specific cocktail of growth factors and culture media. These cultured cells are referred to as ESCs. ESCs can be proliferated to very large numbers in the laboratory. They grow rapidly without any unwanted genetic changes. However, at any point during the expansion, ESCs can be triggered to re-enter their normal developmental programme and differentiate into a range of different cell types and tissues.

The discovery of ESCs was truly remarkable. Scientists found they could essentially suspend the pathway of normal development and differentiation (the inner cell mass cells only exist transiently during normal development), capturing the cells' pluripotent state in the laboratory culture dish. The vast differentiation potential was most strikingly demonstrated when scientists reintroduced mouse ESCs into a different recipient mouse blastocyst. The cells re-entered their normal developmental programme and generated a full range of different cells and tissues of the adult mouse! An overview of this programme was presented in Chapter 2. ESCs can therefore be grown and easily genetically engineered in the laboratory—then used to recreate mice that carry those same genetic modifications. This underpins technologies used to create genetically engineered mice and has led to countless discoveries in biology and medicine. For these astonishing discoveries a Nobel Prize was awarded to Martin Evans for his work on the mouse. In the late 1990s, a similar strategy demonstrated ESCs can also be obtained from human blastocysts (left over from *in vitro* fertilization fertility clinics).

The discovery of human ESCs raised the prospect of creating unlimited quantities of any human cell type in the laboratory culture dish. This has profound implications for fundamental scientific research, as it enables us to generate the 'right' kinds of cells with normal behaviours that can be used as an experimental tool to dissect the function of each of the genes in the human genome. This was a major step forward, compared to working with cancer cell lines, which inevitably have many features that deviate from normal cell behaviour. More information about cell lines can be found in Chapters 2 and 5. ESCs and their differentiated progeny can also be used to model certain genetic diseases. They can also be deployed in chemical or functional genetic screens to find new drugs or drug targets. Finally, certain cell types (e.g. liver and neurons) are especially useful for toxicology studies.

One of the most seductive possibilities is that we could generate specific human cells and use these in new types of transplantation medicine. This is particularly appealing for diseases where a single cell type is lost or damaged. For example, patients with Parkinson's disease would benefit from transplantation of new dopamine-producing neurons; those with type 1 diabetes mellitus (diabetes was discussed in Chapter 5) require functioning pancreatic islet cells that produce insulin; and patients with burns or certain skin disorders require sheets of epidermal cells (see Figure 6.1 to understand the sources and applications of ESCs).

Could we produce all of these easily in the laboratory by differentiated human ESCs? Could we go even further and create organized tissues, rather than just individual cell types? Can complex tissues and rudimentary organs be produced outside a natural embryonic and fetal environment? These are all

Fig. 6.1 Embryonic stem cells are derived from the early blastocyst stage embryo and can be used for multiple applications.

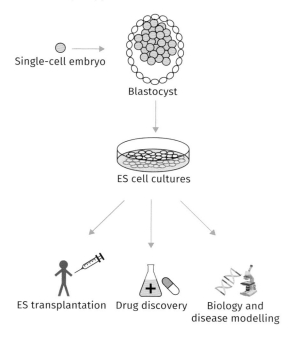

Single-cell embryo

Blastocyst

ES cell cultures

ES transplantation   Drug discovery   Biology and disease modelling

areas of research being actively pursued. The promise is real. And there has been slow and steady progress over the past twenty years. However, there are very real barriers to achieving this goal. The greatest difficulty has been learning how to 'tame' the vast differentiation potential of ESCs to generate fully functional differentiated cells.

Although the pluripotency of ESCs is their remarkable defining feature, this wide differentiation potential can also be problematic: the cells have too many options and can be difficult to control. Steering the various differentiation pathways in a controlled way in the laboratory has been very challenging. The route from the immature pluripotent state to the final fully functioning cell type of interest involves a large series of distinct developmental states. Essentially, the ESCs must be steered along a long and winding journey. We try and mimic the strategies and molecular programmes that are used by the normal embryo to do this. However, in many cases we don't have the right 'map' of the routes for this journey or the molecular signals that can direct them correctly. Finding new ways to control cell differentiation is one of the areas where synthetic biology tools and approaches are proving powerful (see 'Stem cell differentiation'). Put simply: how do we make ESC differentiation faster, simpler, and more reliable?

Tissue stem cells

Tissue stem cells are responsible for renewing and repairing our adult bodies. In stark contrast to ESCs, they are not pluripotent. Instead, these cells produce only the restricted set of cell types needed within the tissue in which they reside; that is, skin stem cells make skin cell types and intestinal stem cells make intestinal cell types. Often, they are referred to as multipotent; that is, being able to generate multiple distinct differentiated cell types. However, there are notable exceptions to this, such as spermatogonial stem cells, which only make a single cell type. Tissue stem cells are often studied in the context of adult tissue physiology and disease. However, it is important to remember that these cells are also present during fetal development and early life, where they generate the vast numbers of cells needed to build the tissues and organs.

One of the best-studied multipotent adult stem cells is the blood stem cell—or haematopoietic stem cell (HSC). Two key features of stem cells have emerged from many years of investigation into the properties of HSCs. First, tissue stem cells reside in a specialized microenvironment—termed a niche—which provides the range of signals needed to sustain self-renewal and limit differentiation. These can be incredibly complex, with a repertoire of distinct growth factors, extracellular matrices, metabolic and hormonal signals, and sometimes neuronal inputs. For HSCs, the niche is located in the bone marrow. This is why bone marrow transplants are often used clinically to treat haematological disorders—the treatment is essentially an adult stem cell transplant that rejuvenates the blood system in transplant recipients.

Second, HSCs do not directly generate their mature progeny (e.g. white blood cells and red blood cells), but instead transit through a series of more restricted progenitors (an immature cell with differentiation potential but limited capacity for self-renewal). An important concept is that of a differentiation hierarchy. The process of differentiation is not reversible and therefore cells gradually lose their potency through development and differentiation. This is also the case for

Fig. 6.2 Stem cells often reside within a specific 'niche' or microenvironment where there are the appropriate signals (growth factors or matrix; curved lines and grey dots, respectively) to sustain their self-renewal. Upon exit from the niche, the cells will progress to a progenitor state and subsequently differentiate.

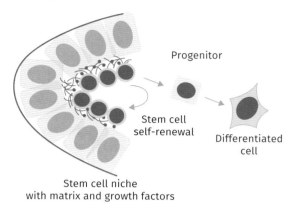

Progenitor

Stem cell
self-renewal

Differentiated
cell

Stem cell niche
with matrix and growth factors

ESC differentiation. It is hard to go backwards along the differentiation path to regain multipotency or pluripotency (see Figure 6.2 to understand the key features of stem cells and progenitors).

Tissue stem cells might seem easier to control compared to the challenges of working with ESCs, as they have more restricted options. However, the major limitations of tissue stem cells are that they are often rare within the tissue and they are difficult to expand in the laboratory. The HSCs highlight this issue. Despite many decades of research, we are still not able to easily expand HSCs in the culture dish—which is why we still need blood donors and bone marrow transplants. It is also important to remember that not all tissues contain stem cells. For example, the adult pancreas and liver don't seem to be maintained and repaired by stem cells, but instead use proliferation of the differentiated cells. Scientists make use of various different experimental techniques (both *in vitro* and *in vivo*) to try to track and explore stem cell behaviour. One emerging realization is that studying tissues under normal homeostasis is very different to that in times of regeneration and repair. When tissues are injured through physical damage, pathogens, or in the context of cancer, different stem cell responses are seen. Some of the 'rules' regarding the hierarchy of differentiation seem to be broken, and progenitors, or even differentiated cells, have been shown to acquire stem cell properties.

Finally, it seems that some of the key features of HSCs do not necessarily hold true in other tissues. Each tissue appears to have acquired distinct stem cell-based strategies for maintaining healthy function. For example, our recently improved knowledge of intestinal stem cells shows clear differences from HSCs. Unlike HSCs, intestinal stem cells are not particularly rare—they make up a large proportion of the tissue. They are also not slow and asymmetrically dividing (features often ascribed to tissue stem cells). Intestinal stem cells are also rather easy to expand in laboratory culture—once scientists realized the key ingredients from the niche and could mimic this in the dish. They form structures in three dimensions called organoids, which continue to grow, sustaining their own niche and showing organized differentiation behaviour that is remarkably similar to their behaviour in the intestinal crypts. These types of organoid culture condition now seem to work well for many types of epithelial stem cells (e.g. pancreas, liver, and oesophagus). There is much still to learn!

In summary, whether using ESCs or tissue stem cells, a major goal is to generate unlimited quantities of desirable human cell types in the laboratory. These would find many applications as an experimental tool to explore basic cell and developmental biology, or for cell transplantation, modelling disease, and use in drug discovery. Alternatively, if we can learn enough about the molecular programmes that control stem cell behaviours in our adult bodies, we might uncover new ways to stimulate repair and rejuvenation of damaged organs. However, a major challenge exists. It has not proven simple to control the differentiation pathways of stem cells in the laboratory.

Stem cell differentiation

Finding efficient and reliable ways of taming the vast differentiation potential of stem cells is one area where the tools and concepts from synthetic biology can help us. Until recently our ability to control stem cell differentiation has been limited to attempts to recapitulate the processes of normal development or regeneration in the laboratory dish. There have been successful examples of this for many different cell types. However, this is often incredibly difficult to achieve as development involves complex interactions and signals, many of which are difficult to recreate and the precious growth factors are often very expensive. So the methods used are difficult and unreliable. It is also slow. Many weeks or months are needed to force cells down their normal developmental programme. Is there a way to bypass the normal rules of development? What are the molecular mechanisms that impose the unique functional and morphological features of a differentiated cell type? What makes a neuron a neuron and not a muscle or skin cell? In the following sections, we discuss the key molecular mechanisms that are responsible for imposing a specific differentiated cell type. A particular class of protein is essential for this: the transcription factors.

Transcription factor 'master regulators' and reprogramming

In the early 1980s, developmental biologists used classical genetic approaches to seek the mechanisms that build the body. By mutating flies, fish, and worms (the model organisms), and looking for offspring in which the embryo had disruption to its normal development, they could try to find 'rules' of development. This approach led to the discovery of many genes important for major events in normal embryonic development such as establishment of the body plan, specifying the lineage identity of early tissue progenitors, and controlling their differentiation. Most of these genes encoded proteins of a class known as transcription factors.

Transcription factors are proteins that bind directly to DNA and alter the levels of the nearby gene expression, by either activating or repressing the activity of RNA polymerase (the enzyme that makes mRNA). There are around 1800 different genes that encode transcription factors in mammals. They fall into a variety of different families that are typically grouped together based on similarities in the three-dimensional structure of their DNA-binding domains. Transcription factors are gene switches. They turn on and off the expression of specific subsets of genes in the genome (see Figure 6.3 to understand how transcription factors work and their modular nature). It is for this reason that they are so fundamental to the control of cell type identity and differentiation.

A neuron differs from a muscle because it has all of the neuron-specific genes turned on (to make, for example, the molecular machinery of the synapse, neurotransmitters, and voltage-gated ion channels), but all of the muscle-specific genes (e.g. muscle-specific myosin and actin) are turned off. The studies of the developmental mutants being investigated in the model experimental organisms told us that if you mutate and destroy the activity of a transcription factor that is important for switching on the genes required by a specific cell or tissue type, then this cell type will not form. An example is the transcription factor Pax6, a transcription factor that is normally active in eye development. Without Pax6 the embryo emerges with no eyes. Remarkably, when researchers activated Pax6 in the wrong region of the embryo, they then saw the formation of eyes at that site! Pax6 is one of many types of master regulator that were identified by such work.

Fig. 6.3 Natural transcription factors are modular with distinct DNA binding and activation domains (a). By tethering activation domains such as VP160 to a DNA-binding dCas9 protein, a synthetic transcription factor emerges that can be placed at any target site in the genome. When near a gene, the RNA polymerase will be recruited and transcription initiated.

(a) *Sequence-specific transcription factor*

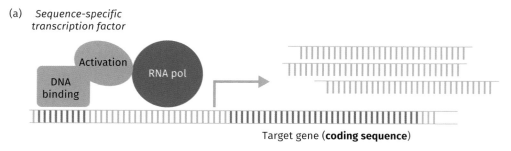

Target gene (**coding sequence**)

(b) *Synthetic transcription factor*

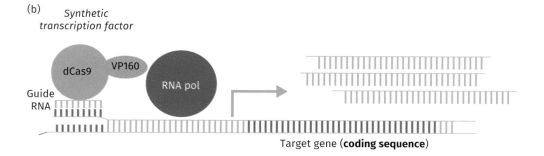

Target gene (**coding sequence**)

Direct reprogramming

The importance of master regulator transcription factors was powerfully demonstrated by scientists studying a muscle master regulator, called MyoD. To test how potent MyoD was, scientists delivered this protein (via a cell-based gene overexpression system) into a non-muscle cell. They chose a cell called a fibroblast, a type of support cell located in the dermis layer of our skin. This is a favourite of scientists studying reprogramming, as it can be easily isolated from embryos or adult skin and is easy to grow and manipulate in laboratory.

Remarkably, when MyoD was delivered to the fibroblasts, they underwent transcriptional reprogramming to acquire muscle cell features. This type of direct reprogramming didn't require cells to reset themselves to a primitive early pluripotent state. Instead, they were forced to switch their identity directly by turning on muscle genes and restricting expression of fibroblast-associated genes. A new gene regulatory network emerged that conferred the cell as a muscle cell type. Since those pioneering studies, a slew of other similar studies have been performed over the past decade, demonstrating with the same strategy (of course, using different master regulators) that fibroblasts can be given new identities; that is, converted to a new different adult cell type.

Reprogramming to pluripotency

If forcing expression of transcription factors can convert one cell type to another, then could it also convert a differentiated cell type to a stem cell? After all, a stem cell is just one specific type of cell type, again defined by the range of genes turned on and off. Experiments involving cloning (making genetically identical copies of the organism; essentially identical twins of different ages) of frogs and sheep gave us reasons to believe this could be possible.

It is now somewhat taken for granted that all cells in our bodies contain the full genome sequence. However, in the 1950s it was still considered possible that cell differentiation could involve a process where the cell loses or actively deletes certain sections of DNA no longer needed for that specific cell type. After all, why would a cell keep the full repertoire of genes for making a muscle if it had decided to be a neuron? However, cloning experiments using nuclear transplantation of an adult frog intestinal cell genome into the single-cell egg suggested this was not the case. The result of these experiments was that a new adult organism can be produced from the nucleus of an adult mature cell type. This demonstrated spectacularly that adult cells do not dispense with any genetic information. Each cell retains the potential to generate an entire organism. Using an alternative strategy, it also was demonstrated that, if adult cells are fused together with ESCs, the resulting hybrid cell was able to adopt ESC identity. Some researchers refer to this change as 'reprogramming'.

It was these experiments that inspired the Japanese scientist Shinya Yamanaka and his team to search for a cocktail for transcription factors that might be able to reprogramme differentiated cells to ESC-like cells. Was there a factor like MyoD that could be the master regulator of ESC identity? Through a systematic filtering and testing of all the transcription factors that are uniquely found in ESCs, they identified a cocktail of four transcription factors—aptly now called Yamanaka factors—that can do this. This was a landmark for the field: cellular alchemy.

Adult specialized cells can now be routinely reprogrammed back to an induced pluripotent cell state (termed iPS cells). This discovery and the associated technologies immediately overcame some of the major difficulties with

Fig. 6.4 Delivery of synthetic transcription factors by transfection can be used to turn on endogenous master regulatory transcription factors, which in turn will alter the cell identity. In this instance, a fibroblast is converted into a muscle cell (myoblast) by activation of *MyoD*.

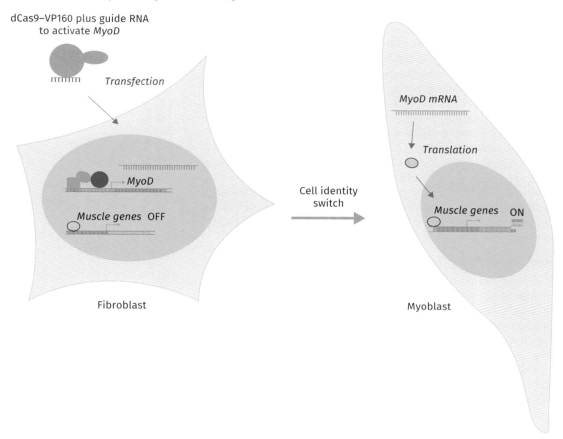

human ESCs, namely that there was less need to work with human embryos from *in vitro* fertilization clinics—simplifying the logistics and removing ethical issues—and also, iPS cells would be genetically matched to the patient/donor, removing the issue of immune rejection that would come with ESC-derived cells. However, the bottleneck in differentiation still remains. More than ever, if we want to exploit the full potential of iPS cells and ESCs we need new and improved strategies to direct their differentiation (Figure 6.4 summarizes some of the best-studied examples of reprogramming).

Synthetic transcription factors

Biochemical studies of transcription factor protein structure identified physically separate protein domains. Some regions were DNA-binding regions, involved in binding specific sites in the genome, while other domains were structurally separate and involved in recruitment of other proteins that trigger or repress transcription (e.g. RNA polymerase, or the chromatin regulatory proteins that package DNA). In other words, transcription factors are modular. This feature is used experimentally to carry out domain-swapping experiments, in which an activator region could be replaced by a repressive region from a different protein. This form of protein engineering can enable new protein activities to be designed, mixing and matching features of interest.

Investigators realized that they could alter the DNA binding domain, particularly those sequences known to interact with the nucleotide sequences in the DNA. This results in a change in the specific DNA sequence to which the transcription factor binds. This led to the emergence of designer or synthetic transcription factors, in which the sequence-specific binding could be specified by design. These were first developed using modified 'zinc-finger' DNA-binding domains (a particular type of natural transcription factor). However, the rules for designing these with correct specificity and activity were not reliable and the protein engineering required was not easy. Another type of transcription factor from plant pathogens, called TAL effectors, emerged which had a remarkably simple set of rules for programming the DNA-binding domain to a sequence of choice. However, TAL effector-based synthetic transcription factors (TALENs) retained some limitations.

Then, over the past decade a new type of system has emerged: CRISPR/Cas9. Few discoveries have had such a rapid and profound impact on the field of molecular and cellular biology as CRISPR! Its main use is as a powerful system for genome editing—the manipulation of gene sequences through inserting, cutting, pasting, or editing a particular target gene (see Chapter 3). However, it has also been repurposed for use as synthetic transcription factors and has quickly become the favoured system for creating synthetic transcription factors.

CRISPR/Cas-based synthetic transcription factors

The Cas9 enzyme in the CRISPR/Cas system is guided by a short RNA to a complementary, matched sequence in the genome. Scientists have re-engineered the Cas9 protein domains, based on knowledge of its structure and the specific nuclease, to remove the DNA cutting activity. A catalytically dead Cas9 (dCas9) emerges that is useful as a foundation for building new synthetic transcription factors. It can easily be targeted to bind specific sequences via the RNA, and the dCas9 can be fused in modular fashion to other effector domains.

Building upon the past knowledge of zinc fingers and TAL effectors, scientists realized that this could probably work well as a synthetic transcription factor if it were tethered to a transcriptional activation domain. A widely used transcriptional activation domain is from the herpes virus and is called VP16. By putting together ten copies of VP16, a much more potent activation domain emerges (termed VP160). The dCas9–VP160 chimeric protein—a fully engineered novel protein—can be guided to any specific site in the genome by conjugating it with a short RNA. The practical consequences of this is that it is now very easy to design, build, and test (the 'mantra' of the synthetic biologist: see Chapter 3) synthetic transcription factors for applications of interest. The 'code' to design a gRNA is very simple and commercial suppliers can make custom gRNAs very quickly. These can be easily delivered by transfection to mammalian cells.

Many laboratories around the world are now exploiting this technology to see how well it works for their gene in their cell type of interest (or even organism of interest, as these can be delivered to single-cell embryos, or via viruses to adult tissues). Proof-of-principle has already been obtained for numerous different genes demonstrating that dCas9–VP160 (or related versions of this) can activate the neighbouring gene. Because of the simplicity of this system, it is clear that a variety of different effector domains can be tested for instead of VP160. These include repressor domains, to turn genes off, and a myriad of chromatin regulatory domains that can alter the histone packaging of a gene, which is often an important element in whether a gene is expressed. We are only at the tip of the iceberg in exploiting this new technology.

Controlling mammalian cell identity and behaviour using synthetic transcription factors

An obvious opportunity now presents itself: can we harness synthetic transcription factors to turn on and off the master regulatory genes in mammalian cells? Can we precisely control their endogenous master regulators and gene regulatory networks, thereby 'forcing' them to a specific cell type or state? The signs so far are extremely encouraging. There has been demonstration for many reprogramming systems, including MyoD or Yamanaka factors being activated in fibroblasts. By activating the endogenous natural master regulators (rather than forcing them by overexpression) we might anticipate a more reliable and robust reprogramming. The correct gene regulatory controls and splicing systems will be operational very quickly.

At present it seems that these synthetic transcription factors do indeed work well, and a major opportunity is that we can deliver different types of regulators at the same time to alter the expression of all the key genes that define a cell type identity. So multiple master regulators for cell type 'A' become activated; and even simultaneously we could repress all the master regulators for cell type 'B'. An example of the success of this approach has come from the laboratory of Charles Gersbach, who successfully activated the neuronal master regulator, called ASCL1, and was thereby able to convert a fibroblast into a neuron. While we are still in the early days of this type of experiment, it looks very likely that these new tools will augment, or perhaps soon replace current cell-differentiation strategies.

We can envisage a time not too far away when mammalian cell gene expression patterns will be engineered by design with exquisite precision. This will have many immediate applications as a useful tool in disease modelling. One example is that the many genes and associated regulatory elements that have been identified and linked to human genetic diseases can be explored more easily. It might soon be possible to engineer stem cells in order to produce cell types that have been engineered with elaborate genetic circuits (encompassing Boolean logic gates, feedback circuits, or toggle switches) that can release therapeutic products, but in a controlled manner with appropriate physiological integration. Perhaps patients will no longer need to inject insulin or therapeutic proteins—instead, there will be cells transplanted with appropriate genetic circuits to provide a more controlled release.

An ability to create specific cell types and test the consequences of gene activation or repression enables us to precisely model disease mechanisms in a very controlled way. Is a particular gene of interest really driving the disease cell pathology? Can we identify key therapeutic targets—genes which when repressed or deleted will lead to normalized cell behaviour? The ability to perform functional experiments in human cells with exquisite genetic precision will therefore have major implications. Moreover, these cells can themselves be used in drug discovery and toxicity testing, including new types of phenotypic reporters that can identify in live cells their status for key cell biological process (e.g. metabolic state, cell cycle, autophagy, and apoptosis). Can we make functional liver cells? Can we make large quantities of HSCs or red blood cells? Can we model genetic diseases precisely and then screen for drugs that 'fix' the disease phenotype?

The road ahead

It is clear that combining our new-found technologies in stem cell biology, reprogramming, and synthetic transcription factors has opened up new ways to explore gene function and regulation, and is helping improve programming and reprogramming of cell type identity. However, thinking further into the future, what other possibilities emerge?

A related opportunity is to use these types of synthetic factors to provide new 'orthogonal' gene circuits (see Chapter 1) and deliver these into mammalian stem cells. The vast amounts of microbial genome sequence data also provide new natural proteins and enzymatic activities that can be integrated with dCas9 to create new types of chimeric proteins with functions not normally seen in mammalian cells. Advances in DNA synthesis enable these to be built cheaply and quickly and so we should see a diverse 'toolbox' of validated synthetic transcription factors and synthetic chromatin regulators that have distinct useful properties and many applications. The possibilities are vast. We can begin to take control of mammalian cell gene expression in ways unthinkable even only a few years ago, creating circuits that express therapeutic proteins of interest, fix disease pathways, or have biosensors that can enable monitoring of tissue health. It is the ability to replace our damaged and disease bodies with laboratory-grown cells and tissues encompassing engineered genetics circuits that will transform medicine in the coming decades.

Related to this is the use of these tools in new types of gene therapy. The major issue with gene therapy has been the inability to get DNA constructs into large numbers of cells safely, without random insertion into the genome. The technologies and approaches that are being developed in gene therapy contexts are also bringing new possibilities when combined with synthetic and stem cell biology. In this scenario, the human cells themselves (not viruses) will be the vehicle for delivering the therapeutic. Then there would be no need for viral delivery and all the associated difficulties. Anti-cancer gene therapies are most likely to emerge, as there have been significant improvements in the delivery with new viral vectors (e.g. adeno-associated viruses have recently been approved).

In conclusion, it should have become clear throughout this chapter that many powerful technologies are converging to lead to new innovative approaches that integrate core methods and principles from synthetic biology, stem cell biology, and gene therapy. The next generations should live longer and healthier lives if these discoveries and tools are effectively translated into new types of therapeutics. The coming decades will be exciting times to be working in these areas!

Chapter summary

- Stem cells are of broadly two categories—pluripotent and multipotent.
- Stem cells can be expanded in the laboratory (e.g. as ESCs, tissue stem cells, and organoids).
- Adult tissues exhibit homeostasis, through repair and regeneration.
- Adult cells can be reprogrammed to pluripotency.

- Reprogramming can be performed directly (rather than via a pluripotent state) using the appropriate master regulator proteins.
- Gene expression can be controlled using synthetic versions of transcription factors and chromatin regulators.
- Future applications of the technologies described in this chapter include the construction of synthetic gene regulatory circuits to control behaviour of human cells, gene therapy for diseases, and the construction of engineered tissues.

Further reading

Church GM, Regis E (2011, reprinted 2014). *Regenesis*. New York: Basic Books.

This book, by one of the world's leading synthetic biologists, provides an overview of the field and speculates intelligently on its implications for future medical technology.

EuroStemCell. https://www.eurostemcell.org.

This site, run by a variety of scientists, is intended to inform European citizens about stem cells.

Gersbach CA (2014). Genome engineering: the next genomic revolution. *Nature Methods* 11, 1009–11.

This is a detailed academic review and commentary, intended for scientifically educated readers.

Yamanaka S, Blau HM (2010). Nuclear reprogramming to a pluripotent state by three approaches. *Nature* 465, 704–12.

This is a detailed academic review and commentary, intended for scientifically educated readers.

Discussion questions

6.1 What criteria does a cell have to meet to be called a stem cell?

6.2 What advantages may there be in controlling stem cell behaviour using synthetic transcription factors instead of natural ones?

7 THE ETHICS OF SYNTHETIC BIOLOGY

David Obree

Learning Objectives

- Describe the importance of ethics in the development of synthetic biology.
- Explain why the ethics of synthetic biology differs from the ethics of analytic biology.
- Identify the key areas of ethical concern in synthetic biology.
- Describe the ethical responsibility of scientists and governments to engage with and inform the public.

The excitement generated by the potential of synthetic biology must be tempered with a degree of caution. As we develop the means to manipulate, enhance, and synthesize biological systems we also develop the means to cause lasting damage, to ourselves, non-human species, and the environment. On the one hand, we want to embrace the beneficial opportunities that the technology offers; on the other, we must be alert to the dark consequences that might prevail. The science, for the time being, is controlled by humans and, while we are an intelligent, inventive, and creative species, we also have an unfortunate and hubristic tendency to destroy the world around us. In short, can humans be trusted to harness the power of synthetic biology wisely and safely?

Awkward questions need to be asked about the purpose and limits of the technology, and hypothetical benefits must be assessed and balanced against possible harms. The potential for deliberate, or accidental, misuse of synthetic biology has to be anticipated and existing beliefs, such as that the natural is always good, need to be challenged. Open discussion involving the public, scientists, governments, and regulators should be encouraged to ensure that safety mechanisms and protective protocols are in place, and that the practical and economic benefits are distributed fairly. It is therefore necessary, and important, for us all to understand and engage with the ethical issues raised by synthetic biology.

Ethics overview

Ethics is the branch of philosophy, sometimes called moral philosophy, that looks at right and wrong, good and bad.

> **Key point**
>
> Historically, **anthropocentric**—that is, centred on human behaviour and inter-action—contemporary ethics has a broader focus which addresses concerns across all the living and geophysical world (Figure 7.1).

How we value non-human species and the environment is central to the discussion. **Instrumental value** is recognizing the usefulness of something to fulfil a goal or purpose, such as animals for meat and wood for shelter. Ecologists and environmentalists, however, argue for **intrinsic value**, that animals and trees have value 'in themselves'. This is a difficult philosophical position to cement as it is only humans that can judge and articulate value, which animals and plants cannot do. Thus advocacy on one level can become stewardship on another, when active measures are taken to look after and protect the non-human world. Public awareness and sensitivity to the plight of animals has increased in recent decades so that vegetarianism and veganism, once viewed as eccentric fads, are now accepted tastes and political positions. Animal supporters who abhor the killing of animals for food, clothes, or product testing are unlikely to support the use of pharming (discussed in Chapter 5). Concerns might be ameliorated if animal welfare is given priority, at least in pharming the animals are not killed, but more fundamentally how can we justify altering an animal's genome for the benefit of humans? By remaining obstinately anthropocentric—always putting humans first—we invite the charge of speciesism.

Fig. 7.1 Historically, anthropocentric, contemporary ethics has a broader focus, which addresses concerns across all the living and geophysical world.

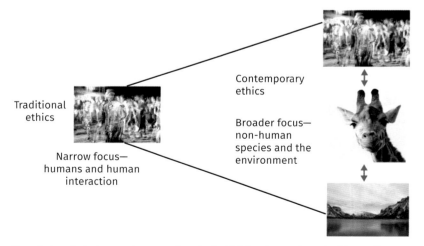

Top photo: photo by goashape on Unsplash. Middle photo: photo by Photo Lily on Unsplash. Bottom photo: photo by Natalie Toombs on Unsplash.

Key point

Central to ethical discourse is (a) who decides what is right and wrong, good or bad; and (b) how decisions are made about what is right and wrong, good or bad.

Sociologists and anthropologists study the **values, norms,** and **relationships** that drive human behaviour; ethicists study not what *has* been done, or what *is* being done, but '*what should be done*'. This is often referred to as the is/ought distinction. Just because something exists, and can be observed, does not necessarily mean it is right. For example, slavery was a 'social norm' until the eighteenth century, and for much of the nineteenth century women were prohibited from attending university. We might argue that the values underpinning ethical decision-making are relative to a time, place, or culture, but we can also make the 'ought' argument that these practices were not right, then, now, or in any circumstance.

 Key point

Ethicists study not what *has* been done, or what *is* being done, but '*what should be done*'.

Throughout history, religious beliefs have influenced values concerning the human body. In Tom Morris's book on the history of heart surgery (*The Matter of the Heart*), he points out that until the mid-twentieth century surgeons were reluctant to operate on the heart, such was the religious and cultural reverence for the beating organ. Further back in time, these beliefs meant that dissection of the human body was entirely prohibited. Today, varying sensibilities about the moral status of human tissue means that legislation on embryo research differs from country to country. As synthetic biology has significance for all of humanity, it would seem desirable to seek a unified, international consensus on the aim and boundaries of the science.

From analysis to synthesis

When Francis Crick and James Watson burst into the Eagle public house in Cambridge, England, on 28 February 1953 to announce that they (from the data of Rosalind Franklin and Maurice Wilkins) had found 'the meaning of life', the ensuing libations were in celebration of the discovery of DNA's structure, a triumph of scientific analysis, not the potential to manipulate DNA itself (Figure 7.2).

Yet one of the scientists whose work Crick and Watson had relied on was to become the most active in raising concerns about the move from analysis to synthesis. Erwin Chargaff, an Austrian biochemist working at Columbia University (New York, USA), had established the equivalence between the bases adenine and guanine, and cytosine and thymine (Chargaff's rule). Chargaff had been inspired by Oswald Avery's work linking DNA to genes and was always conscious of the profound ethical issues at stake. He recognized that the discoveries were leading to a pivotal point in human history, where the science of the knowledge of biological systems (analysis) would become the science to control, change, and build biological systems (synthesis).

Fig. 7.2 The commemorative plaque outside the Eagle public house, Cambridge, England.

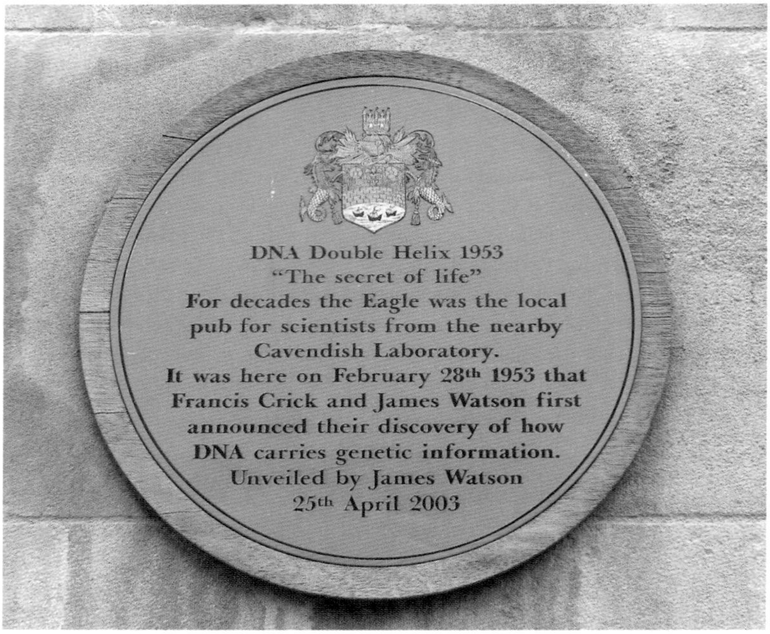
DNA Double Helix 1953
"The secret of life"
For decades the Eagle was the local
pub for scientists from the nearby
Cavendish Laboratory.
It was here on February 28th 1953 that
Francis Crick and James Watson first
announced their discovery of how
DNA carries genetic information.
Unveiled by James Watson
25th April 2003

Source: https://commons.wikimedia.org/wiki/File:220211-030_CPS_(5468852951).jpg Photographer, Chris Sampson, CC-BY-2.0

The ethical and scientific parallels for Chargaff were contained in the simultaneous developments in nuclear physics, where the knowledge of discovery (analysis) had been transmuted into activity (synthesis) with a destructive and unprecedented outcome: the atomic bomb. Identifying that the splitting of the atom and the discovery of DNA both involved 'mistreatment of a nucleus', Chargaff lamented that 'in both cases I have the feeling that science has transgressed a barrier that should have remained inviolate'. His deepest worry was that, while you can stop making bombs, you cannot recall new forms of life once they have been made and set free. Famously irascible, and no doubt irritated by not receiving a Nobel Prize himself, Chargaff tended to dwell on the dangers of synthetic biology instead of the opportunities. However, more recently, and with similar concerns, Friends of the Earth, in collaboration with other environmental groups, produced a manifesto, *The Principles for the Oversight of Synthetic Biology* (2012), calling for strict control of the science and a ban on manipulation of the human genome. We might argue that our existence is due to genetic mutation, not stagnation, and that scientific progress should not be stifled. If genome manipulation is wrong, we must be clear why it is wrong.

The appeal to nature

Key point

The first objection to the emergence of synthetic biology is that it is simply 'not natural'.

Natural is good in many value systems; in philosophical discourse this is called the 'appeal to nature'. Religious systems often refer to nature and natural phenomena as 'God's will'. However, ethicists do not accept this view, since it is the distinction between *is* and *ought* that most concerns them. For example, nature (what *is*) is not all good. There are some terrible and cruel things that occur in nature, natural biological phenomena such as cystic fibrosis, early-onset dementia, Huntingdon's disease, and cancer. It would be perverse to describe these occurrences as good just because they are natural, or that cancer-curing chemotherapy is bad because it is not natural. Indeed, the essence of human and veterinary medicine is to challenge and alter the 'natural' disease process.

> **Key point**
>
> This failure of the appeal to the natural world as the arbiter of good, the failure to link **is** with **ought**, is also called the **naturalistic fallacy**, that not all natural (or observed) things are necessarily good.

In theology, this is also a challenge to the notion that all God's work is for the best. Religion and science share a long history disputing the origins of creation; synthetic biology means this tension has heightened as humans develop the means to rearrange the 'natural', or even become the creators.

Potential benefits, potential harms, and uncertainty

Here is an opportunity for the synthetic biologist to stake a **moral claim**: what if their endeavours could remove the 'natural' bad and replace it with the 'unnatural' good? Such a measurement of removing bad in favour of good is based on **outcomes** and, in ethical terms, is grounded in **consequentialism**.

The potential 'good' outcomes of synthetic biology are boundless. Therapies for debilitating inherited diseases, therapies for cancer, and biogenerated vaccines would all be significant gains for both individuals and the human population (Figure 7.3). Reversing the explosion of type 1 diabetes mellitus utilizing the synthetic biological device discussed in Chapter 1 would bring health and

Fig. 7.3 Balancing harms and benefits.

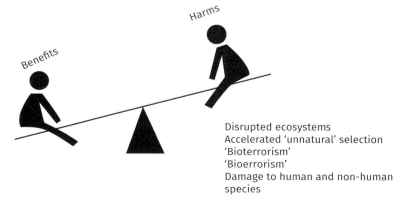

Therapies for debilitating inherited diseases and cancer. New organisms producing bio-fuel to solve the world's energy and pollution crises

Benefits

Harms

Disrupted ecosystems
Accelerated 'unnatural' selection
'Bioterrorism'
'Bioerrorism'
Damage to human and non-human species

well-being to millions of people around the world. Directing and controlling cellular regeneration has obvious potential—imagine the delight of being able to grow a new set of teeth when your old ones have worn out, or being able to grow a new limb following accidental damage, or a new organ when an old one is diseased! Directing and controlling cellular regeneration is explored in Chapter 6.

Concerns that we are tinkering with nature seem insignificant when the benefits are substantial and the harms negligible, but are we doing the opposite of Chargaff by running with the benefits and failing to stop and think about the consequences? In the first instance, should we be thinking more about the range of biotechnologies and identifying the most hazardous? It seems, on the surface, that the dangers of somatic cell manipulation can be contained as the effects are limited to the life of the organism involved. The major concern, and the crucial distinction, comes with germ-line therapies, alterations to the genome that are passed on to subsequent generations. Here we have an escalation of the dangers, as Chargaff pointed out, when it comes to outcomes that are not naturally limited and cannot be reversed. In this context, synthetic biology exits in an **escalating hierarchy**. **Non-heritable gene therapy**, **germ-line therapy**, **enhanced, altered**, and ultimately **new organisms** are steps both in the scientific ladder of achievement and the ethical ladder of concern (Figure 7.4). An understanding of the science and clear definitions are crucial to the ethical discussion.

So, it seems that consequentialism is straightforward—if the benefits exceed the harms, you have a consequentialist 'good'. But what level of harm, or risk of harm, to ourselves, non-human species, and the environment is acceptable? Would harvesting a dozen human neonates as a source of stem cells be acceptable if it helped a thousand old people with Alzheimer's disease? The consequentialist mantras 'the ends justify the means' and 'the greatest good for the greatest number' are also its weaknesses—substantial harm is ignored on the grounds of overall benefit and minorities are overlooked, or simply not given a voice.

 Key point

This raises issues of **justice** for the under-represented in human societies, and for non-human species. What was once called animal rights is now conceived in terms of justice, notions rooted in fairness, and equal opportunity to thrive.

Fig. 7.4 Ethical concerns inevitably exist in an escalating hierarchy. Non-heritable gene therapy, germ-line therapy, enhanced, altered, and ultimately new organisms are steps both in the scientific ladder of achievement and the ethical ladder of concern.

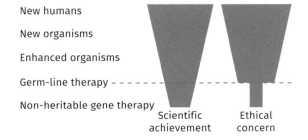

New humans

New organisms

Enhanced organisms

Germ-line therapy - - - - - -

Non-heritable gene therapy

Scientific achievement

Ethical concern

Although traditional science (analysis) is grounded on the concept of observational certainty, we cannot always be sure how new biological systems (created by synthesis) might interact, especially over time. Applying the butterfly effect to delicately balanced ecosystems means that a small genetic change in an organism could have a devastating effect on the viability of that organism, or other species, in years to come. Just as in a game of chess, what seems to be a smart move in the short term might have unintended consequences in the long term. In synthetic biology, those unintended consequences could be checkmate for the human race.

> **Key point**
>
> The worry that technology could fall into the wrong hands and feature in the form of intentional **bioterrorism** is coupled with the parallel fear that technology could escape through simple error, wreaking havoc through unintentional '**bioerrorism**'.

Such fears mirror the worries arising from the misuse of, and accidents with, nuclear energy. In the same way, the dangers of synthetic biology call for international attention, with robust monitoring and policing of the science.

Human enhancement

To many of us, genetic manipulation to remove the disadvantage of disease is morally acceptable, yet artificial enhancement to improve people beyond the norm remains problematic. We accept an athlete who trains hard or is talented by genetic chance but feel uncomfortable with one who takes performance-enhancing drugs or has had their genetics manipulated. The search for human improvement has some merit; after all, we have arrived here through the process of natural selection and the 'survival of the fittest'. However, it is also a flawed aspiration, as demonstrated by the eugenic and Nazi projects of the twentieth century, which defined the ideal human in terms of race and ethnicity. Such racially motivated discrimination violates issues of justice and equality. Here is a possible role for synthetic biology: what if access to the genetic enhancement technologies became available to all, so that there was no privilege? The consequences might be a reduction in inequality with better, happier, longer lives for everyone. A tempting proposition, but also a world of unforeseeable social challenges.

Cloning, the replication of DNA to form an identical organism, is a common science fiction concept heralded as a means to improve the human stock by replicating those with desirable traits. Although brought to reality by Dolly the sheep, the concern about cloning is that, on a large scale, it would end natural selection, the survival of the fittest, and thus evolutionary improvement. Synthetic biology could also interfere with natural selection in the opposite way, by artificially accelerating (controlled) mutations. Although the fittest might still prevail, it would pose a challenge not just to theological and cultural norms, but also to current scientific and secular norms which are based on a longer timescale for (random) mutations to occur.

Anxiety about interfering with nature, and the effect of unbridled science, is not new, it is age-old and prevalent in mythology, literature, film, and television (*Prometheus, Frankenstein, Blade Runner, Ex Machina, Humans, Black Mirror*). Residing in cultural norms and accepted universal beliefs, the anxiety operates at the core of what it means to be human, with Ruud ter Meulen explaining that such narratives function as:

> precautionary fables [. . .] **semiotic** entities embedded with moral messages concerning man's relationship with nature, of playing God, the vices of hubris and the dangers of curiosity and meddling with things we do not fully understand.

That we fear our own death and the end of human existence is an interesting span of individual and communitarian instinct; why should we care for, or have any responsibility for, future generations?

In Ridley Scott's 1982 dystopian science fiction film *Blade Runner* (based on Philip K. Dick's 1968 novel *Do Androids Dream of Electric Sheep?*), a human protagonist (Deckard, played by Harrison Ford) is charged with seeking and destroying a group of errant 'replicants' (Figure 7.5). These genetically created and enhanced humanoids are physically superior to humans, but lack 'human' emotion and have only implanted memories.

The narrative drive of the film is that the 'authentic' or 'native' human race must be protected. Yet it is conceivable that enhanced humanoids could demonstrate greater intelligence, compassion, and reasoning than 'authentic' humans. At what point would enhanced humanoids achieve moral superiority and the moral status of 'authentic' humans diminish? Virtue ethics champions the character of the agent carrying out an act, what happens when 'artificial' virtue triumphs over 'natural' virtue?

Key point

Along with the **is/ought** distinction and the **naturalistic fallacy**, we also have to ask, does nature confer moral superiority?

Fig. 7.5 The replicant-hunting 'Blade Runner' Deckard uses the (fictional) Voight-Kampff machine to test suspected replicants for (human) emotional responses such as dilation of the pupil.

Fig. 7.6 We know that Rachael is a replicant, but is Deckard an (unknowing) replicant too? Are we all just DNA replicants of a sort?

The replicants in *Blade Runner* are ostensibly 'artificial' but then all humans are just products of a chemically activated genetic code, randomly programmed by natural selection (Figure 7.6). We have become comfortable with the science of evolution, but now as we approach the moment of creation, of ourselves by ourselves, we face a new dilemma: if we can become the creators, could others have become creators before us, is this where theology and science begin to converge, or begin to diverge?

The environment

Synthetic biology is not as high in the public and political consciousness as climate change, yet both have the potential for destructive consequences. Politics is essentially societal ethics, what is right and wrong, good or bad for a population, and the same issues of who decides and how decisions are made remain central to the discussion. The global warming debate has been frustrated by a number of factors: a lack of understanding of the science, vested economic and national interests, an absence of the political nerve needed to challenge human consumption, a naive assumption that problems will sort themselves out, and from this heady mix a lack of consensus that means damage to the biosphere continues unabated. The United Nations has been un-united and ineffective in its handling of climate change, an in-collaboration that does not augur well as a model for governing synthetic biology. Perhaps the robust attention given to nuclear safety is a better paradigm; however, the radiation leaks at Three Mile Island, Chernobyl, and Fukushima could all have been prevented, could all have been worse, and were all due to human error and failures in risk assessment.

How can we trust humans to monitor and control the challenges posed by synthetic biology?

Ironically, both nuclear energy and synthetic biology could help to reduce the pace of climate change by providing substitutes for the burning of fossil fuels, if only we could be certain of the safety. Faced with the spectre of global warming, an inevitability, are the risks associated with biofuel derived from biosynthetic algae a gamble worth taking? Already we have biosynthetic crops that increase the efficiency of food production, with the claim that further development could rid the world of famine, but do we want another agricultural revolution and population increase? Over-population is an environmental issue which could be made worse as humans live longer—another potential outcome of synthetic biology—positive for individuals but negative for the biosphere.

By increasing our control over biological systems through synthetic biology we may reinforce our anthropocentric view of the planet, to the ultimate detriment of the non-human world. On the other hand, there could be positive inferences; a deeper reflection of our role in nature might prompt a new sense of responsibility, a new conception of stewardship. In the narrative of the Anthropocene, humans should be the heroes, not the destructive force. Can we learn to match our scientific intelligence with an emotional humility that recognizes the value of the world around us?

Responsibility and safety

We have seen that central to any ethical analysis is how we measure, and balance, good outcomes against bad outcomes. Let us go back to the question of *who* decides what level of bad outcome, or harm, is acceptable. This is not a decision for scientists, clinicians, and governments—that would be paternalism—but something that calls for democratic participation through public consultation and engagement. Transparency, education, and involvement are key factors that underpin the public's trust in science, scientists, and governments, trust that is damaged by secrecy and lack of consultation. Deciding how governments regulate and police the technology is difficult enough on a national level, obtaining international collaboration and agreement a real challenge of diplomacy, but given the dangers, should synthetic biology be given the same level of oversight as nuclear energy? If the dangers are *why* we need to regulate, we also need to ask the following:

> *Who* to regulate? All scientists working in synthetic biology, only those working on genome editing, or only those working on heritable genome-editing?
>
> *What* to regulate? All synthetic biology, or only high-risk synthetic biology?
>
> *How* to regulate? Licensing, registration, inspections, and sanctions for non-compliance?

In 2010, President Obama formed a Presidential Commission to analyse and discuss these issues. The ensuing report, *New Directions: The Ethics of Synthetic Biology and Emerging Technologies*, advocates the concept of 'prudent vigilance' involving concern for (1) **public beneficence**; (2) **responsible stewardship**; (3) **intellectual freedom** and **responsibility**; (4) **democratic** deliberation; and (5) **justice and fairness**. Similar European initiatives, including SYNBIOSAFE and SYBHEL, have reinforced the case for international cooperation along with suggested regulatory frameworks on safety. These initiatives are the first step in creating a consensus of purpose to monitor and control the science and, while regulation is seen as essential, there is nevertheless a caution that it should not be so burdensome as to prevent research altogether.

This deontological approach, creating rules and cultivating a duty among scientists to follow those rules, has potential flaws: who decides that the rules are appropriate and what happens when they conflict? Responsible scientists will fulfil their duty to engage with the policies and protocols but is it naive, cynical, or complacent to believe that someone, somewhere, at some time will be prepared to break the rules? The rogue scientist is a favourite fictional mainstay, but in 2018 we had the story of the Chinese researcher He Jiankui and human twins born with manipulated genomes to protect against HIV. This brings fiction to life and raises another angle of concern, the dangerous use of genome editing for a benefit that could be solved by less risky means. His intervention has been widely condemned, both in his own country and worldwide, but how was he allowed to carry out the procedure in the first place? How can the global community really get organized to police and prevent such activities?

Justice and fairness

Central to the debate surrounding justice and fairness is who should have access to, and who should control, the potential applications of synthetic biology. Justice in health means giving people, and non-human species, the opportunity to thrive and fulfil their potential, defined by the philosopher Norman Daniels as 'normal species functioning'. Illness frustrates this functioning and creates inequalities of opportunity. Will the beneficial therapies arising from synthetic biology be available to everyone, across all nations, or will the daunting costs attract only commercial interest and therefore restrict access only to those in wealthy nations? Contemporary medical ethics has concerned itself with respect for autonomy as its driving principle, a Western zeitgeist formed out of the abuses of paternalism and medical research but propagated by consumerism, anti-communism, and more recently the 'me, me, me' celebrity culture. In this individualistic world, we must ask when does respect for autonomy become respect for selfishness? Individualism obsesses with negative liberty, 'freedom from' (state interference), yet our autonomy also derives from positive liberty due to state intervention, the 'freedom to' that arises from clean water, sanitation, and law enforcement. When the leader of the People's Judean Front (played by John Cleese) asks his followers 'What have the Romans done for us?' in the Monty Python film *Life of Brian* (1979, Terry Jones), the stuttering replies include sanitation, medicine, education, public order, irrigation, the fresh water system, and public health, or as Thomas Hobbes wrote in *Leviathan* (1651), without organized society life would be 'solitary, poor, nasty, brutish and short'. The inference is that we need civilization and society for our autonomy, our positive 'freedoms to'.

Maybe we have become too individualistic and there is a valid and persuasive claim for a more communitarian approach, driven by a desire to reduce inequalities in health but also becoming a necessity due to healthcare inflation and resource constraints. Since inequalities in health can be perpetuated by resource limitation, and with the potential for synthetic biology to offer a practical and economic solution to these woes, there is an economic *and* regulatory argument to locate synthetic biology in the public health sphere.

Here we might consider the differences, and similarities, between synthetic biology and traditional pharmaceutical development, where commercial and national instincts are known to direct and to obstruct access. In pharmaceutical research, innovations aimed at wealthy communities are more likely to be funded than purely humanitarian projects. To fulfil the goal of public beneficence and responsible stewardship, experiments in synthetic biology should avoid exploiting weak socioeconomic groups and non-human species. In short, the benefits and burdens should be distributed fairly within the concept of justice. Existing guidelines on consent, research, and clinical trials may need substantial revision before being applicable to synthetic biology. Accordingly, existing guidelines on animal research will need to be addressed and the boundaries of decency in our use of animals re-examined. The question of what moral justification we have for interfering with non-human species will need careful attention as we explore the impact of synthetic biology on biodiversity, ecosystems, food, and energy supplies.

Issues of ownership will also have to be considered: who will own the intellectual property rights of these new technologies and will the financial benefits be distributed fairly? Will copyright extend to the genetic codes and organisms that are being developed, will living entities be subject to patent? These are further questions not just for scientists, ethicists, and governments but for international lawyers charged with protecting the rights of those concerned. We know that pharmaceutical companies seek to protect and recoup their investment and development costs before their patents run out; is this model sustainable in synthetic biology where, as we learned in Chapter 5, the cost of developing and producing biopharmaceuticals runs into the hundreds of billions? We might prefer that such ethically sensitive developments are funded publicly, but even so, it is likely to remain the preserve of wealthier nations.

Conclusion

Synthetic biology poses both scientific and ethical challenges. As our scientific knowledge develops, our ethical and emotional intelligence will also need to grow. Existential questions about our place in the world will need to be answered and, ultimately, we may have to consider our role in terms of **humanitarian virtue**; what sort of people do we want to be, what relationship do we want with non-human species and the environment? Synthetic biology's potential to create new species but also to wipe out old species by disease *and* enhancement (artificial natural selection) calls for restraint by scientists—not everything that *can* be done *should* be done. Diminishing the risk of a biological Armageddon may not suffice; the risk should be removed altogether—accentuating the positive means eliminating the negative. This demands public recognition and understanding of the issues in order to validate firm monitoring and control of the science through international cooperation and consensus. The test for the human species will be contained in this stewardship.

Chapter summary

- Ethics is the study of right and wrong, good and bad, between humans and between humans, non-human species, and the environment.
- Ethics is concerned with how things should be, not how they are, or have been.
- Central to ethical discourse is (a) who decides what is right and wrong, good or bad; and (b) how decisions are made about what is right and wrong, good or bad.
- Biological synthesis raises many more ethical issues than (traditional) biological analysis.
- The appeal to nature, that natural is good, and artificial is bad, can be challenged.
- The benefits and harms of synthetic biology need to be weighed up and assessed by public engagement and debate.
- Synthetic biology challenges theological, cultural, and secular norms.
- The moral status of the artificial could potentially exceed the natural.
- The unknown and potential power of synthetic biology demands high levels of responsibility from governments and scientists.
- Justice demands that the benefits and burdens of synthetic biology are distributed fairly.

 # Further reading

Chargaff E (1978). Mrs. Partington's Mop, or the Third Face of the Coin. In: *Heraclitean Fire: Sketches from a Life before Nature*, pp. 180–91. New York: The Rockefeller University Press.

The testy molecular biologist's take on his science and the threat to humanity.

Friends of the Earth, CTA, ETC Group (2012). The Principles for the Oversight of Synthetic Biology.

http://www.etcgroup.org/content/principles-oversight-synthetic-biology.

A useful summary of the environmental concerns and precautions relating to synthetic biology.

Kaebnick GE, Murray TH (eds) (2013). *Synthetic Biology and Morality: Artificial Life and the Bounds of Nature* (Basic Bioethics). Cambridge, MA: MIT Press.

A useful textbook with multiple authors offering a range of political and philosophical perspectives on the key issues in synthetic biology.

ter Meulen R, Calladine A (2010). Synthetic biology and human health. Some initial thoughts on the ethical questions and how we ought to approach them. *Rev Derecho Genoma Hum/Law Hum Genome Rev* 3, 119–41.

An overview of the main issues relating to the ethics of synthetic biology by distinguished philosopher Ruud ter Meulen, in collaboration with Alex Calladine.

Presidential Commission for the Study of Bioethical Issues. New Directions: The Ethics of Synthetic Biology and Emerging Technologies.

https://bioethicsarchive.georgetown.edu/pcsbi/sites/default/files/PCSBI-Synthetic-Biology-Report-12.16.10_0.pdf.

The report by the Presidential Commission looking at issues of safety, prudent vigilance, and responsible stewardship.

Sandel MJ (2007). *The Case against Perfection: Ethics in the Age of Genetic Engineering*. Cambridge, MA: The Belknap Press of Harvard University Press.

American philosopher Michael Sandel discusses human enhancement with attention to bionic athletes and designer children. He contrasts the 'old' eugenics of the twentieth century with the potential of 'new' (liberal) eugenics, along with the concept of giftedness, that our genetic makeup is a gift we are obliged to make the most of.

[Further watching] *Blade Runner—The Final Cut* (2007). Ridley Scott (Dir). USA: Warner Brothers.

Classic science fiction exploration of a future dystopia and, by contrast with 'replicants', what it means to be human.

 Discussion questions

7.1   How do the ethical challenges in biological analysis differ from those in biological synthesis?

7.2   Should we protect the natural world by banning synthetic biology?

7.3   How will the concept of 'prudent vigilance' help direct ongoing research?

GLOSSARY

alternative splicing a phenomenon in which **exons** can be skipped during RNA **splicing** to generate alternative forms of a protein.

anoikis a form of cell suicide triggered by cells not receiving signals that would normally come from their correct neighbourhood, including the extracellular matrix with which they should be in contact.

antibiotic selection a method for selecting cells that have been genetically modified successfully among many that have not. In a typical experiment, the researcher arranges that the introduced set of genes results in production of a protein that makes a cell resistant to an otherwise lethal antibiotic, as well as in production of whatever else she wants cells to make. When the antibiotic is applied, only the correctly engineered cells can survive.

antibody format an alternative molecular format of a full-size monoclonal antibody developed in the process of protein engineering (e.g. BiTE®, Cimzia®).

anthropocentric centred/focused on human behaviour/activity.

antigen a toxin or other (usually foreign) substance which induces an immune response in the body, especially the production of antibodies.

auxotrophic a cell that is unable to produce certain metabolites for itself, so is dependent on externally supplied nutrients for survival.

beneficence the ethical principle of doing or promoting good.

bioerrorism the concept that dangerous biological products can be released by (unintentional) error. The word is an informal but widely used play on 'bioterrorism'.

bioinformatics computer science used for analysing **omics** data using quantitative biology techniques and algorithms.

biologic (as a noun) any variety of medicine produced within living organisms rather than by chemical synthesis. Biologics include serums, vaccines, and therapeutic proteins.

bioterrorism the threat or use of dangerous biological products for intentionally malicious purposes.

blastocyst an early stage of embryo formation, in which there is an inner cell mass (that goes on to make the body proper, and a few temporary structures) inside a fluid-filled cavity, surrounded in turn by cells that will make placenta and other temporary structures.

breast duct a tree-like system of tubes within the breast, which carry milk to the nipple.

butterfly effect a term coined by the meteorologist Edward Lorenz to refer to the way in which tiny changes can produce large and effectively unpredictable effects in chaotic systems. It comes from his example of flapping of a butterfly wing disturbing air in a way that multiplies and results in a bigger, more significant event such as a tornado.

Cas9 an RNA-guided endonuclease that is one of the most commonly used effector proteins for CRISPR-mediated DNA editing.

cell line a line (**clone**) of mammalian (or other animal) cells that have been mutated so that they proliferate in culture without limit.

chimaera an organism made from cells of two different origins, for example, a mouse made from an embryo to which ESCs from a different mouse have been added.

chimaeric protein a protein made up of parts of other proteins (because it is made from a gene that consists of parts of other genes): synthetic biologists often make chimaeric proteins as a way of combining functions normally found in separate natural proteins.

chromosome a complex of DNA, and proteins around which the DNA is wrapped, and more proteins that hold the DNA–protein coil together into a coherent body. The genomes of mammals exist in chromosomes.

clone a set of genetically identical individuals (cells or organisms), or any individual member of that set.

consequentialism the philosophy that the outcomes of an act determine the rightness or wrongness of that act.

CRISPR a term for a method of gene editing based on bacterial RNA-guided endonucleases such as Cas9 to modify DNA. The acronym comes from the bacterial system in which these nucleases were discovered—clustered regularly interspaced short palindromic repeats.

cytoplasm the part of the inside of a cell that is not in an organelle. Organelles are in the cytoplasm.

deontology an act-based theory of ethics relying on a duty to follow a set of rules.

designer nucleases DNA-cutting enzymes, now widely used as a genome editing tool in synthetic biology. Specifically, they can cut double-stranded DNA at user-defined target sites in the genome (genes). Designer nucleases include CRISPR/Cas9, ZFNs, and TALENs.

differentiation literally becoming different; in development, cells changing gene expression to take on a (more) mature state.

differentiation hierarchy the order of differentiation events that restricts the potency of stem cells and their differentiating daughters.

DNA sequencing a biochemical technique for determining the precise order or sequence of nucleotides (A, T, G, C) in individual genes or entire genomes.

dopamine 3,4-dihydroxyphenethylamine, a neurotransmitter.

electroporation a transfection method that uses an electric field applied across the cell membrane to increase cell permeability to allow DNA to enter the cell.

embryonic stem (ES) cell a cell line derived from an early embryo, that is still pluripotent (can generate all cells of the body).

emergence the creation of high-level features of a system from the action of low-level elements that do not themselves have those features. For example, the creation of sand dunes by the action of wind and sand, neither of which contain 'dune-ness' themselves.

endocytosis a natural cellular behaviour in which small zones of plasma membrane bud off inwards to form a vesicle in the cytoplasm, consisting of a membrane sac containing a sample of the liquid that was immediately outside the plasma membrane before endocytosis occurred.

enhancer a gene control element of mammalian cells that is located far from the transcription start site. Proteins binding at enhancers cooperate with those binding at the promoter to control expression of the gene.

epigenetic any influence on gene expression that operates at a level above the sequence of bases in the gene. Transcription factors, chromatin modification complexes, and signalling systems are all examples of epigenetic factors.

epitope a structural motif in a protein or antigen that is recognized by an antibody.

ethics the (moral) philosophy of right and wrong, good and bad.

eugenics the study or belief in human improvement by selective control of breeding. Positive eugenics seeks to improve the human species by encouraging those with 'positive' traits (physical ability, high IQ) to breed; negative eugenics seeks to improve the human species by preventing those with 'negative' traits (physical disability, low IQ) from breeding.

eukaryote a cell with a true nucleus. Animals, plants, and fungi are all eukaryotes, as are many single-celled organisms, but bacteria are not.

exon a protein-coding part of a gene: the primary transcript of most mammalian genes will contain alternating exons and **introns** and the introns will be **spliced** out to leave just the exons joined to one another to make a continuous protein-coding sequence.

feedback a control mechanism whereby the result of some process is fed back to control the process (as when, for example, a heating system measures the temperature of a room to decide how much power should be applied to room heaters).

frameshift mutations genetic mutations caused by the insertion or deletion (indels) of nucleotides in a DNA (or RNA) sequence—putting out the normal reading (codon) frame.

free energy the difference between the energy in a system in its current state and the minimum energy of the system; equivalently, the maximum amount of work that can be extracted from a system.

functional motif an arrangement of just a few elements in a genetic or cellular network, that performs a particular function (e.g. decision-making), the arrangement appearing in many different networks to perform that same function.

gene amplification an increase in the number of copies of a gene. This is often induced by **chromosomal** rearrangements.

gene knockout (KO) an experimental genetic technique (often using **designer nucleases**), designed to disable (i.e. knock out) the normal function of a gene in a cell (*in vitro* transgenics) or in a living organism (*in vivo* transgenics). Gene KO allows us to study gene expression, regulation, and function in synthetic biology.

gene therapy manipulation of the genome with the aim of eliminating the genetic defect responsible for a specific genetic disorder.

genetic code the 'look-up table' between triplets of bases in mRNA and the amino acid that they will cause to be incorporated into a protein being translated. Warning: 'genetic code' is often misused in general language (newspapers, etc.) as a synonym for genome. The definition given here is the only correct one for scientific use.

genome editing the intentional insertion, deletion, or modification of a selected DNA sequence at a specific site in the genome.

germ line the population of cells in the body that can give rise to sperms or eggs. This population is set aside very early in development.

glycosylation the process of adding carbohydrate chains to (some of) the amino acid side chains of a protein.

Golgi apparatus a cellular organelle through which most proteins destined for secretion or incorporation into the plasma membrane pass, and where **glycosylation** takes place.

gRNA (guide RNA) an RNA guide, acting as a locator—used to target Cas9 to a particular complementary location on DNA.

haematopoietic stem cell a stem cell that can make the various cells of the blood.

Hayflick limit the limit to the number of times natural tissue cells can divide.

HeLa a cell line made from the cervical cancer of the African American woman Henrietta Lacks, who died of her disease in 1951.

homologous recombination a somewhat complicated process, which takes place naturally but that can be made more probable with some technical tricks, in which two pieces of DNA carrying the same sequence can be swapped. In the context of synthetic biology, homologous recombination is often used to insert a synthetic genetic element at a precise location in a chromosome (the chromosome has the homologous sequence uninterrupted; the engineered DNA has the same sequence, but interrupted by the piece of new genetic material to be inserted into the chromosome).

hormone a molecule that can mediate long-range signalling between cells that produce it and cells that detect it. Most hormones travel throughout the body in the blood.

housekeeping gene a gene expressed in all cell types, typically because it is needed for basic cellular metabolism that all cells need, whatever their specialized function.

hysteresis the name given to the behaviour of a system in which the dose–response curve follows one line when moving from low dose to high, but another line when moving from high dose to low.

indel a mutation caused by insertion ('in-') or deletion ('-del') of bases from DNA.

iPS cell (induced pluripotent stem cell) a cell in a pluripotent state, that has been created by altering gene expression in a more differentiated cell.

intron a non-coding 'interruption' in the coding region of a gene that has to be spliced out before the transcript becomes mRNA capable of being translated.

in vitro outside the body, in a culture vessel, test-tube, etc.

in vivo in a living body (as distinct from in cell or tissue culture).

is/ought distinction the idea that, just because something is done, exists, or can be observed, does not necessarily mean it is right.

justice theory the moral philosophy concerned with fairness and equality of opportunity; it can include the distribution of, and access to, resources. Also, principles about how benefits and burdens should be shared.

lipofection transfection of a cell with DNA complexed with a lipid to aid crossing of the plasma membrane.

liposome a tiny bubble-like shell of phospholipids, the constituents of normal biological membranes. Experimenters often load compounds into liposomes to deliver them to cells, and can modify the surfaces of liposomes so that they interact only with specific types of cell.

lymph node an organ of the immune system, into which tissue fluid (lymph) flows on its way back to the blood system. Lymph nodes are the main place where immune cells that detect evidence of infections can initiate a defensive response.

master regulator a gene, the expression of which causes cells to make large-scale changes in their gene expression to become a different cell type.

metastasis the spread of cancer cells from their site of origin to other parts of the body.

monoclonal antibody (MAb) an **antigen**-binding protein derived from a single cell, which binds a unique target molecule.

monoclonal cell line a population of cells which have arisen from a single progenitor through repeated cell divisions. Theoretically, all these cells should be genetically identical.

multipotent (of a stem cell) able to make a number of different cell types, but not all. This contrasts with pluripotent.

naturalistic fallacy the error of believing that because something is natural, it must be good.

niche a specialized microenvironment.

nucleofection a variant of electroporation-based transfection that allows DNA to be transferred directly to a cell's nucleus.

nucleus an organelle that contains the **chromosomes**, and transcription apparatus and spliceosomes of mammalian cells.

omics technologies an informal term, derived from the suffix -omic, for a group of methods that are applied to all items in very large biological sets. These include genomics (all genes in the genome), transcriptomics (all transcribed genes in a sample), proteomics (protein content of a sample), lipidomics (lipid species in a sample), and metabolomics (small-molecule metabolites in a sample). Such methods can be used separately, or in combination, to study and connect events occurring at the level of DNA and RNA, with changes in cellular metabolism and protein expression. All omics methods utilize advanced bioinformatics.

organ a physically distinct assembly of **tissues** dedicated to performing overall functions. An organ is a level of organization between tissue and organism.

organelle a compartment within a cell specialized for a particular function. Most (but not all) organelles are bounded by lipid membranes that separate their contents from the bulk cytoplasm.

organoid a three-dimensional structure, formed in tissue culture, that includes at least two tissues of an organ and that is organized in an organ-like way (the criteria for 'organ-like' vary with the application).

paternalism the doctrine that governments or those in authority 'know best' and can overrule an individual's autonomy.

pattern a non-random arrangement of two or more states (colours, lights, cell states, etc.) in space or time.

pharming an informal but widely used word for the process of producing pharmaceutical products within livestock or crops.

phenotype the physical expression of a genotype, determined by the genotype and environmental influences acting together.

phosphatidyl serine a membrane phospholipid, normally confined to the inner leaflet of the plasma membrane by the enzyme flippase, but allowed to enter the outer leaflet in cells dying by apoptosis.

porphyria a metabolic disease resulting from inherited deficiencies in haem synthesis.

post-translational modification an alteration to the amino acid sequence or amino acid side chains of a protein that is carried out following translation.

prokaryote an organism with no true nucleus: bacteria are prokaryotes, but yeasts and mammalian cells are not.

promoter an element just upstream of the transcription start site of a gene at which, in the presence of appropriate transcription factors, RNA polymerase is activated so that it begins to transcribe the gene.

random mutagenesis early attempts at genetic modification using chemical- or radiation-induced methods to randomly mutate genes. This process proved imprecise and inefficient, as researchers need to discard many mutants in the search for the traits they desire.

receptor a biomolecule, usually a protein, that can bind to a signalling molecule and initiate some action in response. Often, receptors are carried on the plasma membrane, where they can bind extracellular ligands and initiate further signalling events within the cell.

regenerative medicine medical interventions intended either to make an injured or missing tissue regenerate, or to replace it with a working version.

restriction enzymes (also called **restriction endonucleases**) protein enzymes which recognize a specific sequence in DNA, and then cut each strand of the sugar–phosphate backbone at, or near, that site (called a restriction site). Restriction enzymes were among the first tools used in molecular genetics, to manipulate components and structures of the genome.

safe harbour site a location on a chromosome in which a transgene may be inserted with only a very low risk of it being shut down by epigenetic chromatin modification, and only a very low risk of it being regulated by nearby genetic elements of the host.

self-renewal the division of a population of stem cells in a way that ensures that on average half the daughters of the division are themselves stem cells, thus maintaining the stem cell population as well as creating daughters to differentiate.

serum half-life the length of time it takes for the amount of a substance introduced into blood (e.g. by injection) sample to be halved.

side effect an effect of a drug secondary to the one intended.

signalling molecule any molecule that can mediate signalling between cells. Signalling molecules include both long-range hormones and short-range signalling proteins.

speciesism the term coined by British philosopher, psychologist, and animal welfare activist

Ricard D. Ryder in the 1970s. Subsequently popularized by the American philosopher Peter Singer to describe discrimination against other species but can also refer to the creation of a hierarchy between species such that, for example, mice are preferred to chimpanzees for experimentation.

spliceosome the complex of protein and small RNA that **splices introns** out of transcripts in the **nucleus**.

splicing (of RNA) the stage of RNA processing that involves removal of **introns** and joining of the **exons** that flank them.

stable cell line a population of cells that maintains its behaviour for a prolonged period of time.

stable transfection transfection of a cell with DNA in a way that ensures that the DNA becomes integrated into the genome as a new permanent feature. Typically, this process is inefficient and involves careful selection of the cells that are stably transfected, to reject all those merely transiently transfected.

stem cell one of a population of cells that can renew themselves and give rise to (usually more than one type of) differentiated progeny. Stem cells are responsible for replacing worn out or damaged cells in many tissues of the adult.

stewardship the concept of (responsible) guidance and control, often in relation to natural resources, the environment, and non-human species.

synapse a specialized site of communication from one neuron to another, or a neuron to a muscle (or, in the case of an immunological synapse, between specific cells of the immune system).

TALENs transcription activator-like effector nucleases are restriction endonucleases (enzymes) specifically engineered to cut a desired sequence of DNA (similar to the ZFN method)

telomerase an enzyme that builds up the telomeres (ends) of chromosomes; without this enzyme, telomeres become shorter with every cell division.

tissue an assembly of cells that carry out particular functions. A tissue is a level of organization between cells and **organs**.

tissue-specific promoter a promoter that is only transcribed within certain **tissues** of the body, where the right combination of transcription factors is present.

tissue stem cell a stem cell in a tissue which can make (some of) the cell types in that tissue, but not other cell types of the body: it is multipotent, not pluripotent.

transfection deliberate transfer of exogenous non-food material (typically DNA) into a cell.

transgene a gene introduced into a cell (usually a genome) from some other organism. In common (but incorrect) use, 'transgene' can also refer to an entirely synthetic gene that has never existed in another organism.

transient transfection transfection of cells with DNA, which does not integrate into the genome. Because the DNA is not in the genome, it will not be replicated when cells divide, and so will eventually be diluted out and lost. This contrasts with stable transfection.

transposons also called 'jumping genes'—segments of DNA that can move around the genome. Important in genome regulation, genetics, and evolution.

transcription factor a factor (typically a protein or a complex of proteins) that controls transcription of a gene by binding to its promoter or enhancer.

transient expression expression of a gene from a plasmid that has not integrated into the genome. Plasmids are degraded in cells or diluted out by cell division, so transient express only persists for a few days.

virtue ethics a character-based theory of ethics whereby good actions arise from individuals with virtuous character traits such as compassion, honesty, and courage.

virtuous cycle a cycle of actions in which each leads to an improvement of the others.

vitamin one of a class of small organic molecules that are essential to the health of an organism. Some are made by the organism, some can be obtained only from the diet (e.g. humans cannot make their own vitamin C).

ZFNs zinc-finger nucleases are engineered restriction enzymes which allow site-specific genome engineering.

zygote a cell formed by the union of two gametes (sperm and egg). The zygote is the simplest form an individual mammal ever has.

BIBLIOGRAPHY

Chapter 1

Xie M, Ye H, Wang H, Charpin-El Hamri G, Lormeau C, Saxena P, et al. (2016) β-cell-mimetic designer cells provide closed-loop glycemic control. *Science* 354, 1296–301.

Chapter 2

Harima Y, Takashima Y, Ueda Y, Ohtsuka T, Kageyama R (2013). Accelerating the tempo of the segmentation clock by reducing the number of introns in the Hes7 gene. *Cell Rep* 3, 1–7.

Takashima Y, Ohtsuka T, González A, Miyachi H, Kageyama R (2011). Intronic delay is essential for oscillatory expression in the segmentation clock. *Proc Natl Acad Sci U S A* 108, 3300–5.

Chapter 3

Hutchison CA 3rd, Chuang RY, Noskov VN, Assad-Garcia N, Deerinck TJ, Ellisman MH, et al. (2016). Design and synthesis of a minimal bacterial genome. *Science* 351, aad6253.

Merkert S, Martin U (2016). Targeted genome engineering using designer nucleases: state of the art and practical guidance for application in human pluripotent stem cells. *Stem Cell Res* 16, 377–86.

Chapter 4

Cachat E, Liu W, Martin KC, Yuan X, Yin H, Hohenstein P, Davies JA (2016). 2- and 3-dimensional synthetic large-scale de novo patterning by mammalian cells through phase separation. *Sci Rep* 6, 20664.

Kramer BP, Fussenegger M (2005). Hysteresis in a synthetic mammalian gene network. *Proc Natl Acad Sci USA* 102, 9517–22.

Onuma H, Komatsu T, Arita M, Hanaoka K, Ueno T, Terai T, et al. (2014). Rapidly rendering cells phagocytic through a cell surface display technique and concurrent Rac activation. *Sci Signal* 7, rs4.

Pascoli V, Terrier J, Hiver A, Lüscher C (2015). Sufficiency of mesolimbic dopamine neuron stimulation for the progression to addiction. *Neuron* 88, 1054–66.

Yagi H, Tan W, Dillenburg-Pilla P, Armando S, Amornphimoltham P, Simaan M, et al. (2011). A synthetic biology approach reveals a CXCR4-G13-Rho signaling axis driving trans-endothelial migration of metastatic breast cancer cells. *Sci Signal* 4, ra60.

Chapter 5

Abdullah MA, Ur Rahmah A, Sinskey AJ, Rha CK (2008). Cell engineering and molecular pharming for biopharmaceuticals. *Open Med Chem J* 2, 49–61.

Al-Sawaf O, Fischer K, Engelke A, Pflug N, Hallek M, Goede V (2017). Obinutuzumab in chronic lymphocytic leukemia: design, development and place in therapy. *Drug des Devel Ther* 11, 295–304.

Donner A (2013). Synthetic influenza seeds. https://www.nature.com/scibx/journal/v6/n21/pdf/scibx.2013.509.pdf.

Dumont J, Euwart D, Mei B, Estes S, Kshirsagar R (2016). Human cell lines for biopharmaceutical manufacturing: history, status, and future perspectives. *Crit Rev Biotechnol* 36, 1110–22.

Gaidukov L, Wroblewska L, Teague B, Nelson T, Zhang X, Liu Y, et al. (2018). A multi-landing pad DNA integration platform for mammalian cell engineering. *Nucleic Acids Res* 46, 4072–86.

Kemmer C, Gitzinger M, Daoud-El Baba M, Djonov V, Stelling J, Fussenegger M (2010). Self-sufficient control of urate homeostasis in mice by a synthetic circuit. *Nat Biotechnol* 28, 355–60.

Manini I, Trombetta CM, Lazzeri G, Pozzi T, Rossi S, Montomoli E (2017). Egg-independent influenza vaccines and vaccine candidates. *Vaccines (Basel)* 5, E18.

Ronda C, Pedersen LE, Hansen HG, Kallehauge TB, Betenbaugh MJ, Nielsen AT,

Kildegaard HF (2014). Accelerating genome editing in CHO cells using CRISPR Cas9 and CRISPy, a web-based target finding tool. *Biotechnol Bioeng* 111, 1604–16.

Ruan J, Jie XR, Yanru C-T, Kui L (2017). Genome editing in livestock: are we ready for a revolution in animal breeding industry? *Transgenic Res* 26, 715–26.

Toussaint C, Henry O, Durocher Y (2016). Metabolic engineering of CHO cells to alter lactate metabolism during fed-batch cultures. *J Biotechnol* 217, 122–31.

Ye H, Xie M, Xue S, Charpin-El Hamri G, Yin J, Zulewski H, Fussenegger M (2017). Self-adjusting synthetic gene circuit for correcting insulin resistance. *Nat Biomed Eng* 1, 0005.

Young Rojahn S (2013). Synthetic biology could speed flu vaccine production. *MIT Technology Review*. https://www.technologyreview.com/s/514661/synthetic-biology-could-speed-flu-vaccine-production/

Chapter 6

Grath A, Dai G (2019). Direct cell reprogramming for tissue engineering and regenerative medicine. *J Biol Eng* 13, 14.

Marchisio MA, Huang Z (2017). CRISPR-Cas type II-based synthetic biology applications in eukaryotic cells. *RNA Biol* 14, 1286–93.

Slack JMW (2018). What is a stem cell? *Wiley Interdiscip Rev Dev Biol* 15, e323.

Chapter 7

Chargaff E (1978). *Heraclitean Fire: Sketches from a Life before Nature*. New York: The Rockefeller University Press.

Friends of the Earth, CTA, ETC Group (2012). The principles for the oversight of synthetic biology. http://www.etcgroup.org/content/principles-oversight-synthetic-biology.

Kaebnick GE, Murray TH (2013). *Synthetic Biology and Morality: Artificial Life and the Bounds of Nature* (Basic Bioethics). Cambridge, MA: MIT Press.

ter Meulen, R, Calladine A (2010). Synthetic biology and human health. Some initial thoughts on the ethical questions and how we ought to approach them. *Rev Derecho Genoma Hum/Law Hum Genome Rev* 3, 119–41.

Morris T (2017). *The Matter of the Heart: A History of the Heart in Eleven Operations*. London: The Bodley Head.

Presidential Commission for the Study of Bioethical Issues. New Directions: The Ethics of Synthetic Biology and Emerging Technologies. https://bioethicsarchive.georgetown.edu/pcsbi/sites/default/files/PCSBI-Synthetic-Biology-Report-12.16.10_0.pdf.

Sandel MJ (2007). *The Case against Perfection: Ethics in the Age of Genetic Engineering*. Cambridge, MA: The Belknap Press of Harvard University Press.

INDEX

A

Adalimumab 68
Addiction 51–3
Agriculture 78–81
Alleles 17
Alternative splicing 19
Analytic approaches to biology 1, 99
Anthropocentricism 98
Arthritis 68
Asexual reproduction 17
Avery, Oswald 99

B

Bacteria 6, 15, 16, 26
BioBricks 39, 41
Bioerrorism 103
Bioinformatics 46
Biologics 7, 67
Biosimilars 67
Bioterrorism 103
Blade Runner 104–5
Blood–brain barrier 29
Blood vessels 3, 5

C

Cancer 7
Cells 2, 22, 26, 28, 60, 72–3, 75,
    85–7, 91–2
Chargaff, Erwin 99
Chemical sensing 8
Chimaera 10
CHO cells 72–3, 75
Chromatin 21–2, 24
Chromosomes 16, 17, 21
Clone 8
Combinatorial control 21
Communication 3, 5, 24
Compartmentalization 3, 23
Consequentialism 101
CRISPR 37–9, 41, 45, 46–7, 93–4
Culture of cells 26
Cytoplasm 22

D

Decision-making 58
Design, build, test cycle 33
Development 4
Diabetes melitus 7, 9, 12, 77–8
Dick, Philip K 104
Differentiation 3, 8, 85, 89
Disease 7
DNA repair 45
DNA sequencing 34–6, 45
DNA synthesis 39–40

Dolly (the sheep) 25
Dopamine 51
Dystrophin 19

E

Electroporation 42, 74
Embryonic stem cells 10, 85–7, 91
Emergence 5, 11
Encapsulation 9
Endoplasmic reticulum 23
Energy landscape 64
Engineering cycle *see* 'Design, build,
    test cycle'
Enhancers 20
Environment 11, 105–6
Epigenetic control 19
ES cells *see* Embryonic stem cells
Ethics 97–110
Eugenics 103–4
Eukaryotes 15, 22
Evolution 57
Exons 18
Extracellular matrix 27

F

Fat 5
Feedback 5, 24, 57–60, 77
Fingerprints 5
Fluorescent proteins 26
FosB gene 51
Functional motifs 58

G

Genes 1, 5, 10, 37, 74
Gene amplification 74
Gene deletion 1
Gene modification 5, 10, 37
Genetic code 16
Genome editing 37
Germ line 4
Glycosylation *see* Post-translational
    modifications
Golgi apparatus 23, 60
Good manufacturing practice 73
Gout 14, 77–9
Growth 5
Gurdon, John 25
Gutenberg, Johannes 34

H

Haematopoietic stem cells 87
Hayflick limit 28
Hek293 cells 73
HeLa cells 60
Hes7 gene 19

Hierarchy of organization 2
Homeothermy 2
Hormones 3, 5, 27
Housekeeping genes 23
Human enhancement 103–5
Hypothesis testing 8
Hysteresis 58

I

Induced pluripotent stem cells 91–2
Immortomouse 28
Immunity 6, 26, 29–30
Inflammation 6, 7, 26, 29, 68
Insulin 7
Introns 18
iPS cells *see* Induced pluripotent
    stem cells
Is/ought distinction 99

J

Justice 102, 107–8

L

Lipofection 42
Livestock 80–1
Lysosomes 22

M

Macrophages 29
Maintenance 5
Mammals (general properties) 2, 16
Mammary glands 2
Markers 26
McClintock, Barbara 39
Medicine 7, 84–95
Milk 25
Mitochondria 22
Modelling 8, 21, 56–7
Monoclonal antibodies 68, 74
Muscle 5
Mutagenesis 37

N

Naturalistic fallacy 101
Nervous system 3
Neutrophil 26, 29
Niche 87–8
Noise immunity 58
Nucleofection 74
Nucleosomes 22
Nucleus 22

O

Oligonucleotide 40
Optogenetics 52–3
Organelles 22

Organoids 88–9
Organs 3

P

Pattern formation 61–4
Phagocytosis 60–1
Pharming 79–80
Phase separation 62
Pluripotency 4, 91–2
Porphyria 12
Post-translational modifications 72, 76
Printing 34
Production of proteins by cells (industrial) 69–2
Promoters 20
Protein folding 7

R

Rapamycin 60
RASSLs 55
Recombination 17, 38, 44
Responsibility 106–7
Restriction enzymes 37
Ribonucleic acid 16, 18

Ribosomes 16
RNA *see* Ribonucleic acid

S

Safety 28, 106–7
Scale 2
Selection 9, 75
Senescence 28
Sex 17
Speciesism 98
Splicing 18, 19
Spliceosomes 18
Stem cells 4, 84–9
Switches 11

T

TALENs 37
Telomere 28
Timing 19
Tissues 3, 87–9
Tissue stem cells 87–9
Transcription 16, 18, 20
Transcription factors 20–1, 24–5, 89–94
Transfection 42–4

Translation 16, 18
Transposons 39

V

Vaccination 29, 66, 68–9
Vascular endothelial growth factor 5
VEGF *see* Vascular endothelial growth factor
Vertebrates 2
Virtue 108
Virtuous cycles 50
Viruses 28, 68–9
Vitamins 12

W

Waddington, Conrad 19
Wilmut, Ian 25

Y

Yamanaka factors 91–2, 94
Yamanake, Shinya 91
Yeast 15, 16, 26, 41

Z

Zinc-finger nucleases 37
Zygote 4